Applications of Ultrasonic Technology in Biomedical Sciences

Applications of Ultrasonic Technology in Biomedical Sciences

Guest Editors

Constanza Rubio Michavila
Sergio Castiñeira-Ibáñez
Daniel Tarrazó-Serrano

MDPI

Basel • Beijing • Wuhan • Barcelona • Belgrade • Novi Sad • Cluj • Manchester

Guest Editors
Constanza Rubio Michavila
Centro de Tecnologías Físicas
Universitat Politècnica
de València
Valencia
Spain

Sergio Castiñeira-Ibáñez
Centro de Tecnologías Físicas
Universitat Politècnica
de València
Valencia
Spain

Daniel Tarrazó-Serrano
Centro de Tecnologías Físicas
Universitat Politècnica
de València
Valencia
Spain

Editorial Office
MDPI AG
Grosspeteranlage 5
4052 Basel, Switzerland

This is a reprint of the Special Issue, published open access by the journal *Applied Sciences* (ISSN 2076-3417), freely accessible at: https://www.mdpi.com/journal/applsci/special_issues/4HRV492AH9.

For citation purposes, cite each article independently as indicated on the article page online and as indicated below:

Lastname, A.A.; Lastname, B.B. Article Title. *Journal Name* **Year**, *Volume Number*, Page Range.

ISBN 978-3-7258-8303-5 (Hbk)
ISBN 978-3-7258-8304-2 (PDF)
https://doi.org/10.3390/books978-3-7258-8304-2

Cover image courtesy of Constanza Rubio Michavila

Contents

About the Editors

Constanza Rubio Michavila

Constanza Rubio Michavila (Castelló de la Plana, Spain) received an MS Physics degree (1990) from the Universitat de València, and a PhD in Applied Physics (1995) from the Universitat Politècnica de València. She is a Full Professor in the Applied Physics Department at the Universitat Politècnica de València and the author of more than 60 research articles published in journals. She has participated in more than 20 public and transfer research projects, and has also contributed to numerous conference papers and book chapters. She is the author of three patents developed at the UPV and actively participates in developing educational improvements and innovative material for the subjects she teaches. She also studies the behavior of waves in the optical range and its influence on monitoring of the environment and crops through Remote Sensing, allowing for more sustainable treatments and policies throughout the entire value chain. From the academic point of view, she is the Director of the School of Agricultural and Environment.

Sergio Castiñeira-Ibàñez

Sergio Castiñeira-Ibàñez (Mislata, Spain) holds a degree (Universitat de València) and PhD in Physical Sciences (Universitat Politècnica de València), and is an Associate Professor in the Department of Applied Physics at the Universitat Politècnica de València. He also worked for an extended period of his academic and teaching career at the Universitat de València. His research focuses on studying the physical properties of periodic systems applied to acoustic waves and lenses for ultrasound transmission, which is carried out at the Centro de Tecnologías Físicas: Acústica, Materiales y Astrofísica, part of the Universitat Politècnica de València. He has participated in more than ten competitive research projects in his field, contributing to the advancement of knowledge and innovation in applied physics. Throughout his research career, he has authored over thirty articles published in JCR-indexed journals, reflecting a broad dissemination of his scientific results. He is also the co-author of more than thirty papers presented at international conferences, several of which were invited contributions. He has co-authored two patents and contributed two chapters in technology books.

Daniel Tarrazó-Serrano

Daniel Tarrazó-Serrano (València, Spain) graduated in 2012 with a B.Sc. in Telecommunications Engineering, with a specialization in sound and images, and in 2013 with an M.Sc. in Acoustics from the Universitat Politècnica de València. He obtained two awards while he was studying: an Academic Award in his second course, and an Excellence Student Award for obtaining the best results. In 2014, he completed his second B.Sc. in Telecommunications Systems, and obtained an M.Sc. in Biomedical Engineering from the Universitat de València and Universitat Politècnica de València in 2016. He started his Ph.D. in 2017 as a senior researcher at the Physics Technologies Center. His line of research focuses on applied acoustics, particularly ultrasonics. He was a collaborating professor in the Applied Physics Department at the Universitat Politècnica de València and a visiting researcher at the School of Non-Destructive Testing and Security (National Research Tomsk Polytechnic University), located in Siberia, Russia. In 2020, he obtained the title of Doctor Cum Laude with a special mention awarded for his extraordinary thesis developed as part of the "Technologies for Health and Welfare" Ph.D. program, "Design, modelling, characterization and implementation of acoustic lenses for modulation of ultrasound beams."

Currently, he is a Lecturer and Professor in the Department of Applied Physics at the Universitat Politècnica de València (Spain). He actively participates in multidisciplinary R&D (in agronomy, environmental acoustics, ultrasound, and numerical modeling), teaching research, public R&D, and transfer projects associated with companies. He has published scientific articles, book chapters, and congress papers in R&D and educational publications, and participates in several university educational innovation projects.

Preface

We are delighted to present *Applications of Ultrasonic Technology in Biomedical Sciences*, a Special Issue reprint dedicated to the latest vanguard in ultrasonic sensing and imaging. Ultrasound continues to redefine biomedical science, serving as an indispensable, non-invasive tool that bridges the gap between tissue characterization, functional diagnostics, and advanced therapeutics.

Inside this volume, readers will find nine curated papers, including eight innovative research articles and one deep-dive review, that capture the breadth of modern biomedical ultrasonics. From clinical diagnostics and preclinical models to quantitative imaging and focused ultrasound therapies, these works showcase cutting-edge developments in imaging methodologies and signal processing.

Crucially, this collection reflects the vibrant, multidisciplinary intersection of acoustic physics, engineering, and frontline clinical practice. By focusing on quantitative, reproducible, and highly targeted techniques, these contributions lay the groundwork for next-generation diagnostics and personalized patient care.

Whether you are a researcher, clinician, or engineer in medical imaging, we trust that this compilation will be an invaluable resource. We sincerely thank the authors for sharing their expertise, and the reviewers and editorial staff for their dedication to bringing this Special Issue to fruition. We hope that these pages inspire new questions and accelerate further breakthroughs in the field.

Constanza Rubio Michavila, Sergio Castiñeira-Ibáñez, and Daniel Tarrazó-Serrano
Guest Editors

applied sciences

MDPI

Editorial

Applications of Ultrasonic Technology in Biomedical Sciences

Constanza Rubio, Sergio Castiñeira-Ibáñez * and Daniel Tarrazó-Serrano

Centro de Tecnologías Físicas, Universitat Politècnica de València, Camí de Vera s/n, 46022 Valencia, Spain; crubiom@fis.upv.es (C.R.); dtarrazo@fis.upv.es (D.T.-S.)
* Correspondence: sercasib@fis.upv.es

1. Introduction

Ultrasonic sensing and imaging technologies play a pivotal role in contemporary biomedical science due to their non-invasive nature, real-time capability, and compatibility with a wide range of biological tissues. Ultrasound consists of mechanical waves propagating at frequencies above the audible range, whose interaction with soft tissues and biological structures provides valuable information on their mechanical, structural, and functional properties. These features have positioned diagnostic ultrasound as one of the most widely used imaging modalities in clinical practice, with established applications in cardiology, neurology, oncology, musculoskeletal assessment, and veterinary medicine [1,2].

From a physical standpoint, ultrasonic wave propagation in biological media is governed by acoustic and elastic wave phenomena, including frequency-dependent attenuation, scattering, and dispersion, which are strongly influenced by tissue microstructure. In this context, quantitative ultrasound techniques have emerged to move beyond qualitative B-mode interpretation by estimating intrinsic tissue parameters such as attenuation coefficients, backscatter statistics, or scatterer distributions. These approaches enable objective and reproducible tissue characterization and have demonstrated particular potential for detecting subtle pathological alterations that may remain undetected using conventional imaging alone [3,4].

Significant advances in ultrasound acquisition and image formation have been reported in recent years. Ultrafast imaging based on plane wave transmission and advanced beamforming strategies has enabled high-temporal-resolution imaging while preserving spatial quality [5]. Alongside this, the development of large-scale and sparse array architectures has improved spatial resolution and near-field performance while mitigating hardware and channel-count limitations, which is especially relevant in biomedical and preclinical imaging scenarios [6,7].

Beyond imaging, ultrasound has become an essential tool for functional and physiological assessment. Doppler-based methods and related ultrasound biomarkers are widely used to evaluate blood flow dynamics, vascular compliance, and hemodynamic responses under different physiological or pharmacological conditions [8]. Likewise, neuromuscular and peripheral nerve ultrasound has gained increasing clinical relevance as a complementary diagnostic modality, providing detailed anatomical information that enhances the accuracy of neuromuscular disorder assessment [9].

Ultrasound is also playing an expanding role in experimental and preclinical research, particularly in oncology and therapeutic applications. Focused Ultrasound (FUS) enables localized energy delivery through thermal or mechanical mechanisms while preserving surrounding tissues, and its use in neuro-oncology, including pediatric brain tumors, is an active area of investigation [10,11]. Furthermore, advances in acoustic field manipulation have fostered renewed interest in acoustic lenses for biomedical applications. Planar Fresnel-type acoustic lenses, including MRI-compatible designs, have demonstrated the ability to achieve spatial focusing and beam shaping without complex electronic control, offering promising solutions for image enhancement, therapeutic targeting, and hybrid imaging environments [12].

Altogether, the continued convergence of acoustic physics, quantitative imaging methodologies, advanced hardware design, and clinically driven applications reflects the dynamic evolution of ultrasonic technologies in biomedicine. The contributions collected in this Special Issue exemplify this multidisciplinary trend, spanning methodological innovation, experimental validation, and applied clinical research, and point toward future ultrasound systems that are increasingly quantitative, adaptive, and application-specific.

2. An Overview of Published Articles

This Special Issue brings together nine high-quality contributions addressing recent advances in ultrasonic sensing and imaging, with a particular emphasis on biomedical and clinical applications. Throughout this section, the numbers in parentheses (e.g., (1), (2), etc.) refer to the individual articles included in this issue, as listed in the Section "List of Contributions". The collection comprises eight original research articles and one comprehensive review paper. For clarity, the contributions are discussed within thematic categories reflecting their main methodological and application domains, while preserving the order in which they appear in the list of contributions.

2.1. Clinical and Functional Diagnostic Ultrasound

Choi et al. (1) investigated the effects of varying caffeine dosages and consumption timings on cerebral vascular and cognitive functions in healthy adults using diagnostic ultrasound. By monitoring carotid artery blood flow, blood pressure, and oxygen saturation, the authors demonstrated that a moderate caffeine intake induces measurable hemodynamic changes. In particular, a dose of 200 mg produced more pronounced increases in systolic blood pressure and oxygenation compared to 100 mg, while moderate caffeine consumption was associated with reduced cognitive response times. The study highlights the sensitivity of ultrasound to functional physiological variations and underscores its potential for non-invasive cerebral assessment, while also emphasizing the importance of defining safe consumption guidelines to maximize stimulatory benefits without compromising vascular health.

Jožef and Podnar (3) examined the role of peripheral nerve ultrasonography as a complementary diagnostic tool in patients with different forms of polyneuropathy. In this retrospective study, ultrasound imaging provided non-invasive anatomical information that complemented conventional electrophysiological testing, particularly by revealing nerve enlargement patterns in demyelinating conditions such as chronic inflammatory demyelinating polyradiculoneuropathy. The results showed that nerve ultrasound yielded clinically relevant additional information in approximately half of the analyzed cases, either by supporting initial diagnostic hypotheses or by identifying focal neuropathies not detected through standard methods. These findings reinforce the role of ultrasonography as a sensitive and practical technique that improves diagnostic accuracy and aids in distinguishing between hereditary and acquired neuropathies.

Esterellas-Sánchez et al. (4) reported a prospective study evaluating the impact of left ventricular ejection fraction on the multimodal assessment of congestion in patients with acute heart failure. By integrating clinical examination, biochemical biomarkers such as NT-proBNP and CA125, and the Venous Excess Ultrasound Score (VExUS), the authors compared different heart failure phenotypes. Although patients with reduced ejection frac-

https://doi.org/10.3390/app16115175

tion presented higher biomarker levels and greater fluid accumulation, venous congestion grades assessed using the VExUS protocol were comparable across all groups. The findings demonstrate that VExUS constitutes an objective and reproducible ultrasound-based metric for quantifying venous congestion independently of systolic function, supporting its use in standardized diagnostic workflows and in guiding personalized decongestive therapy in hospital settings.

2.2. *Ultrasound-Based Experimental and Preclinical Studies*

Ricci et al. (2) investigated the effects of Low-Intensity Continuous Ultrasound (LICU) on three-dimensional PANC-1 pancreatic cancer spheroids as a preclinical model to assess ultrasound-induced viability and structural changes. Using artificial vision tools and confocal microscopy, the authors demonstrated that an excitation frequency of 3.5 MHz was particularly effective in inducing cellular death and reducing spheroid volume. A key finding of the study is that LICU treatment minimized spheroid disaggregation, a critical factor in reducing the risk of metastasis during therapeutic intervention. By proposing a precise methodological framework to optimize ultrasound parameters, this work highlights the potential of non-invasive, non-thermal ultrasonic stimulation as an alternative or complementary strategy to conventional oncological treatments and validates the relevance of three-dimensional tumor models for investigating mechanically induced cellular responses.

Pérez-Marín and Quevedo-Sánchez (8) conducted a comparative study of testicular echotexture and scrotal thermography in beef bulls before and after electroejaculation to optimize reproductive assessment protocols. The results showed a significant increase in testicular echogenicity following semen collection, accompanied by a reduction in scrotal surface temperature of approximately 3 °C, reflecting functional tissue responses to the procedure. The authors also provide methodological guidelines for ultrasound analysis, demonstrating that the evaluation of multiple small regions of interest within the testicular parenchyma yields more accurate and reproducible information than global assessments of the entire testis. Although no direct correlation was found between imaging-derived parameters and semen quality, the study emphasizes the complementary diagnostic value of ultrasonography and thermography for detecting subclinical abnormalities and testicular asymmetries, supporting their role in comprehensive veterinary reproductive evaluation.

2.3. *Quantitative Ultrasound, Advanced Imaging, and Anatomical Assessments*

Elvira et al. (5) proposed an advanced quantitative ultrasound framework aimed at improving biological tissue characterization beyond conventional B-scan imaging. The methodology integrates envelope statistics, modeled through the homodyned K-distribution, with spectral analysis to enable the simultaneous estimation of attenuation coefficients and scatterer clustering parameters. Validation was carried out using polyvinyl alcohol phantoms loaded with different particle concentrations designed to mimic the acoustic properties of real biological tissues. The results demonstrate that biparametric imaging significantly enhances the identification of tissue structures and pathological features compared to traditional single-parameter approaches, allowing a more detailed representation of tissue physical properties. This quantitative strategy represents a promising avenue for improving objective diagnosis and for supporting data-driven and artificial intelligence–assisted medical imaging workflows.

Martínez-Graullera et al. (6) investigated the optimization of volumetric ultrasound imaging using large-scale two-dimensional sparse arrays, with a particular focus on preclinical cerebrovascular studies in small-animal models. The authors addressed key technical challenges associated with conventional large 2D arrays, including tissue attenuation and beam steering limitations imposed by element size. To overcome these constraints, they

https://doi.org/10.3390/app16115175

proposed an aperture diversity–based strategy combined with nonlinear processing techniques, replacing conventional beam tilting with sparse aperture configurations. Through numerical simulations and tailored apodization schemes, the proposed approach achieved enhanced spatial resolution and image contrast while preserving high acquisition speed. The methodology effectively mitigated imaging artifacts and optimized the use of electronic channels, demonstrating its suitability for high-fidelity three-dimensional reconstructions in complex biological environments.

Miguel-Pérez et al. (7) conducted an extensive anatomical, ultrasound, and histological investigation of the palmaris longus muscle, a forearm structure characterized by high anatomical variability and clinical relevance in reconstructive surgery. The coordinated use of ultrasonography, anatomical dissections, and histological analysis across 72 human upper limbs allowed for a precise assessment of muscle presence, dimensions, and morphological variants. The results confirmed that ultrasound provides high accuracy in identifying anatomical variations or the absence of the muscle, which is critical for preoperative planning of tendon graft procedures. Additionally, the study detailed the spatial relationship between the palmaris longus tendon and the median nerve, helping to minimize surgical complications. Overall, the findings underscore the reliability of ultrasound as a non-invasive diagnostic tool and its value in optimizing surgical decision-making without compromising neurological function.

2.4. Review and Emerging Therapeutic Applications

Kleinknecht et al. (9) present a comprehensive review on focused ultrasound (FUS) and its emerging therapeutic applications in pediatric brain tumors, with particular emphasis on Low-Intensity Focused Ultrasound (LIFU) as a non-invasive treatment strategy. The article analyzes the potential of LIFU to address diffuse midline gliomas, a group of pediatric brain tumors historically associated with poor prognosis and limited therapeutic options. One of the key challenges highlighted is the presence of the blood–brain barrier (BBB), which restricts the delivery of most pharmacological agents to the central nervous system. In this context, LIFU emerges as a promising technique capable of transiently and locally disrupting this barrier, thereby enhancing drug penetration into tumor tissue.

The review summarizes preclinical evidence demonstrating that LIFU can significantly increase the efficacy of chemotherapy, immunotherapy, and radiotherapy by improving the delivery of therapeutic agents. Furthermore, the authors discuss ongoing early-phase clinical trials that assess the safety and feasibility of these approaches in pediatric patients, often employing advanced magnetic resonance imaging–guided systems. While preliminary results are encouraging, the review also identifies key technical and clinical challenges, including the complexity of pediatric skull anatomy and the need for standardized treatment protocols. Overall, this contribution establishes a critical reference framework that highlights the transformative potential of focused ultrasound technologies for improving the prognosis of children with aggressive brain tumors and outlines future directions for research and clinical translation in pediatric neuro-oncology.

3. Conclusions

This Special Issue, focused on advances in ultrasonic sensing and imaging for biomedical and clinical applications, provides a comprehensive and up-to-date snapshot of current research trends in the field. The nine contributions included in this collection reflect both methodological innovation and applied relevance, highlighting the continued evolution of diagnostic ultrasound as a key non-invasive tool in modern medicine. Collectively, the published articles report significant progress in areas such as advanced imaging strategies, including plane wave imaging and large-scale sparse arrays, quantitative ultrasound

https://doi.org/10.3390/app16115175

techniques for tissue characterization, and multimodal approaches combining envelope statistics, spectral analysis, and complementary diagnostic information. In addition, several contributions address clinically oriented studies, demonstrating the applicability of ultrasound in neuromuscular, cardiovascular, oncological, and neurological assessments, as well as its role in experimental and preclinical contexts. The thematic diversity of this Special Issue is consistent with the scope outlined in the "Overview of Published Articles" section and underscores the versatility of ultrasonic methods in different biomedical scenarios. From fundamental methodological developments to clinically driven investigations, these contributions collectively illustrate the growing maturity of ultrasonic sensing technologies and their capacity to provide reliable, quantitative, and clinically meaningful information.

In general, this Special Issue highlights the ongoing convergence of technological innovation, signal and image processing advances, and clinical application in the field of ultrasound. We believe that the nine contributions presented here offer valuable information for researchers, clinicians, and engineers working in ultrasonic sensing and imaging. Looking forward, the next great frontier in biomedical ultrasound will involve standardizing complex therapeutic protocols, overcoming anatomical barriers, and simultaneously leveraging quantitative multiparametric data to integrate artificial intelligence into personalized clinical workflows.

The Guest Editors would like to thank all authors for their high-quality contributions and their commitment throughout the review and revision process, which has been essential to the success of this Special Issue.

Conflicts of Interest: The authors declare no conflicts of interest.

List of Contributions:

The articles published in this Special Issue offer a broad and up-to-date overview of recent advances in ultrasonic sensing and imaging, with a strong emphasis on biomedical and clinical applications. The contributions span from methodological developments in plane wave imaging, sparse and large-scale arrays, and tissue characterization techniques, to diagnostic ultrasound studies addressing neuromuscular, cardiovascular, oncological, and neurological conditions. Together, these works illustrate the growing versatility of ultrasound as a non-invasive, multimodal tool for both experimental research and clinical practice. The following list summarizes the contributions included in this Special Issue.

1. Choi, M.-K.; Ahn, H.-S.; Kim, D.-E.; Lee, D.-S.; Park, C.-S.; Kang, C.-K. Effects of Varying Caffeine Dosages and Consumption Timings on Cerebral Vascular and Cognitive Functions: A Diagnostic Ultrasound Study. *Appl. Sci.* **2025**, *15*, 1703. https://doi.org/10.3390/app15041703.
2. Ricci, M.; Dimitri, M.; Serio, M.; Corvi, A. Morphological Analysis of US Treated PANC-1 Spheroids. *Appl. Sci.* **2025**, *15*, 1707. https://doi.org/10.3390/app15041707.
3. Jožef, M.; Podnar, S. The Role of Ultrasonography in the Diagnostic Evaluation of Patients with Polyneuropathy. *Appl. Sci.* **2025**, *15*, 6137. https://doi.org/10.3390/app15116137.
4. Esterellas-Sánchez, L.K.; Campos-Sáenz de Santamaría, A.; Albines Fiestas, Z.S.; Crespo-Aznárez, S.; Sánchez-Marteles, M.; Garcés-Horna, V.; Alcaine-Otín, A.; Giménez-López, I.; Rubio-Gracia, J. How Does Left Ventricular Ejection Fraction Affect the Multimodal Assessment of Congestion in Patients with Acute Heart Failure? Results from a Prospective Study. *Appl. Sci.* **2025**, *15*, 8157. https://doi.org/10.3390/app15158157.
5. Elvira, L.; de León, C.; Durán, C.; Ibáñez, A.; Parrilla, M.; Martínez-Graullera, Ó. Tissue Characterization by Ultrasound: Linking Envelope Statistics with Spectral Analysis for Simultaneous Attenuation Coefficient and Scatterer Clustering Quantification. *Appl. Sci.* **2025**, *15*, 9924. https://doi.org/10.3390/app15189924.
6. Martínez-Graullera, Ó.; Camacho, J.; Huecas, J.; Cosarinsky, G.; Elvira, L.; Parrilla, M. Plane Wave Imaging with Large-Scale 2D Sparse Arrays: A Method for Near-Field Enhancement via Aperture Diversity. *Appl. Sci.* **2025**, *15*, 10934. https://doi.org/10.3390/app152010934.

7. Miguel-Pérez, M.; Muñiz Delgado, A.; Ortiz-Miguel, S.; Lopez-Prades, S.; Ortiz-Sagristà, J.C.; Miralles, C.; Ruiz-Roig, N.; Möller, I.; Martinoli, C.; Pérez-Bellmunt, A. Palmaris Longus Muscle and Its Variations: Ultrasound, Anatomical and Histological Study with Clinical and Surgical Applications. *Appl. Sci.* **2026**, *16*, 817. https://doi.org/10.3390/app16020817.
8. Pérez-Marín, C.C.; Quevedo-Sánchez, L. Comparative Testicular Echotexture and Scrotal Thermography Before and After Electroejaculation in Beef Bulls. *Appl. Sci.* **2026**, *16*, 2780. https://doi.org/10.3390/app16062780.
9. Kleinknecht, S.; Fox, K.; Tsitsos, F.; Zacharoulis, S. Focused Ultrasound (FUS) and Pediatric Brain Tumors: Current Status and Future Directions. *Appl. Sci.* **2025**, *15*, 11322. https://doi.org/10.3390/app152111322.

References

1. Szabo, T.L. *Diagnostic Ultrasound Imaging: Inside Out*, 2nd ed.; Academic Press: Waltham, MA, USA, 2014.
2. Bushberg, J.T.; Seibert, J.A.; Leidholdt, E.M.; Boone, J.M. *The Essential Physics of Medical Imaging*, 3rd ed.; Lippincott Williams & Wilkins: Philadelphia, PA, USA, 2018.
3. Cloutier, G.; Destrempes, F.; Yu, F.T.H.; Tang, A. Quantitative Ultrasound Imaging of Soft Biological Tissues: A Primer for Radiologists and Medical Physicists. *Insights Imaging* **2021**, *12*, 127. [CrossRef] [PubMed]
4. Mamou, J.; Oelze, M.L., Eds. *Quantitative Ultrasound in Soft Tissues*; Springer International Publishing: Cham, Switzerland, 2015.
5. Tanter, M.; Fink, M. Ultrafast imaging in biomedical ultrasound. *IEEE Trans. Ultrason. Ferroelectr. Freq. Control* **2014**, *61*, 102–119. [CrossRef] [PubMed]
6. Drinkwater, B.W.; Wilcox, P.D. Ultrasonic arrays for non-destructive evaluation: A review. *NDT E Int.* **2012**, *48*, 1–17. [CrossRef]
7. Ramalli, A.; Boni, E.; Roux, E.; Liebgott, H.; Tortoli, P. Design, Implementation, and Medical Applications of 2-D Ultrasound Sparse Arrays. *IEEE Trans. Ultrason. Ferroelectr. Freq. Control* **2022**, *69*, 2739–2753. [CrossRef] [PubMed]
8. Hoskins, P.R.; Martin, K.; Thrush, A. (Eds.) *Diagnostic Ultrasound, Third Edition: Physics and Equipment*, 3rd ed.; CRC Press: Boca Raton, FL, USA, 2019. [CrossRef]
9. Mah, J.K.; van Alfen, N. Neuromuscular Ultrasound: Clinical Applications and Diagnostic Value. *Can. J. Neurol. Sci.* **2018**, *45*, 605–619. [CrossRef]
10. Kennedy, J.E. High-intensity focused ultrasound in the treatment of solid tumours. *Nat. Rev. Cancer* **2005**, *5*, 321–327. [CrossRef] [PubMed]
11. Labib, S.; Bright, R.K.; Liu, J. Focused Ultrasound in Cancer Immunotherapy: A Review of Mechanisms and Applications. *Ultrasound Med. Biol.* **2025**, *51*, 1–14. [CrossRef] [PubMed]
12. Tarrazó-Serrano, D.; Castiñeira-Ibáñez, S.; Sánchez Aparisi, E.; Uris Martínez, A.; Rubio Michavila, C. MRI Compatible Planar Material Acoustic Lenses. *Appl. Sci.* **2018**, *8*, 2634. [CrossRef]

https://doi.org/10.3390/app16115175

applied sciences

MDPI

Article

Comparative Testicular Echotexture and Scrotal Thermography Before and After Electroejaculation in Beef Bulls

Carlos C. Pérez-Marín * and Luis Quevedo-Sánchez

Department of Animal Medicine and Surgery, Campus of Rabanales, University of Cordoba, 14071 Cordoba, Spain; luisquesan@hotmail.com
* Correspondence: pv2pemac@uco.es; Tel.: +34-957218716

Abstract

Testicular echotexture and scrotal thermography were evaluated in beef bulls before and after electroejaculation (EE) to assess their response to semen collection and to standardize echotexture assessment methodology. Twelve Limousin bulls (12 months to 5 years of age) were included in the study. Semen analysis revealed that 41.6% of the bulls exhibited low semen quality. Testicular echotexture values recorded before ejaculation were significantly lower ($p < 0.01$) than those obtained after semen collection, and differences between testes were observed in some bulls, suggesting reduced reproductive efficiency. For methodological standardization, echotexture values obtained from a single large region of interest were compared with those from six smaller parenchymal regions; the larger region yielded lower pixel intensity values. Echotexture did not differ among imaging planes (proximal, middle, and distal), although probe–testis distance significantly affected measurements. Thermographic analysis showed that the proximal scrotal region was approximately 4 °C warmer than the distal region, and both regions exhibited a temperature decrease of approximately 3 °C following ejaculation. No correlations were identified between semen quality parameters and imaging variables. In conclusion, testicular echotexture increased, whereas scrotal surface temperature decreased after ejaculation. Although ultrasonography and thermography were not associated with semen quality, they provide complementary information for the detection of subclinical testicular alterations. Further studies are warranted to determine whether disparities between testes may serve as indicators of subfertility in bulls.

Keywords: ultrasound; thermal; echotexture; testis; ejaculation; bulls

1. Introduction

Reproductive efficiency is a key determinant of productivity in cattle farming, both in dairy and beef systems, as fertility directly affects herd profitability [1,2]. Although extensive research has focused on female reproductive performance, the contribution of the male has historically received less attention [1]. Nevertheless, bull fertility exerts a disproportionate influence on herd reproductive outcomes: whereas the reproductive failure of a cow results in the loss of her specific production cycle and associated maintenance costs, a single bull's subfertility can compromise the reproductive success of an entire herd [3]. Consequently, the development of accurate, efficient, and practical methods for evaluating male reproductive performance is essential for sustainable herd management.

Ultrasonography is a rapid, non-invasive diagnostic technique widely used for the routine evaluation of both external and internal reproductive organs in bulls [4,5]. It has proven particularly valuable for the assessment of testicular pathology, as it provides detailed information on tissue architecture and structural integrity. Ultrasonographic images are composed of pixels, each representing a specific tissue density displayed on a grayscale [6]. This technique offers several advantages, including repeatability, safety, and the absence of negative effects on reproductive potential [7,8]. Moreover, ultrasonography enables objective morphometric and textural analyses, facilitating the detection of subclinical or structural abnormalities that may not be apparent during manual palpation [9–11]. In this context, echotexture analysis of testicular ultrasonograms provides a quantitative approach that reduces subjectivity and improves diagnostic consistency [12–14].

Infrared thermography has emerged as a complementary, non-invasive diagnostic tool in veterinary medicine, allowing the assessment of surface temperature patterns and the identification of physiological or pathological changes [15–17]. This technique measures infrared radiation emitted from the body surface and converts it into thermal images that visualize temperature gradients [18,19]. Because thermography does not require physical contact, it is particularly suitable for stress-free temperature assessment in livestock [11,20,21]. Early studies characterized normal scrotal temperature distribution in clinically healthy bulls and compared these patterns with those observed in animals presenting reproductive disorders [22]. Subsequent research demonstrated that environmental factors, such as ambient temperature and humidity, significantly influence scrotal surface temperature measurements [23,24]. Under normal conditions, testicular temperature is maintained approximately 4–5 °C below core body temperature to ensure optimal spermatogenesis [25,26]. Disruption of scrotal thermoregulation—whether due to heat stress, alterations in insulation, or testicular pathology—can impair spermatogenesis and ultimately reduce semen quality [27,28].

Given the increasing interest in precision livestock technologies, the combined use of ultrasonography and thermography may enhance the assessment of bull fertility by providing complementary, non-invasive information on testicular structure and function. In the present study, testicular ultrasonography and scrotal thermography were performed before and after semen collection by electroejaculation (EE). The objectives were: (1) to evaluate changes in testicular echotexture and scrotal surface temperature following semen collection, and (2) to standardize the methodological approach for testicular echotexture assessment by evaluating how technical variations in probe positioning and image analysis criteria affect data consistency.

2. Materials and Methods

2.1. Animals, Management, and Location

The study was conducted in twelve Limousin bulls aged between 12 months and 5 years, all of which had unknown reproductive histories. The animals were managed under extensive grazing conditions on two farms located in Cordoba (Spain) and received supplemental feeding during the summer months. The trial was carried out during the breeding season (November to February).

All semen collection procedures complied with European Union legislation on animal experimentation (Directive 2010/63/EU), as transposed into Spanish law (RD 53/2013), and were part of the farms' routine reproductive health evaluations. The experimental protocol was reviewed and approved by the Animal Experimentation Ethics Committee of the University of Cordoba (approval no. 2022PI/20).

2.2. Experimental Design

On day 0, each bull was restrained in a chute for testicular ultrasonographic examination and scrotal circumference (SC) measurement. This was followed by semen collection

https://doi.org/10.3390/app16062780

by electroejaculation (EE) and an immediate post-ejaculation ultrasonographic assessment. Two days later, each bull underwent scrotal thermography, followed by a second semen collection and a post-ejaculation thermographic evaluation.

Only semen samples obtained on each sampling day were analyzed in the laboratory for motility, morphology, and sperm concentration. Bulls with acceptable semen quality were included in comparative analyses; when differences between collection days were observed, the optimal ejaculate—defined as the sample exhibiting the highest motility and morphological values—was selected for analysis.

A schematic representation of the experimental design is shown in Figure 1.

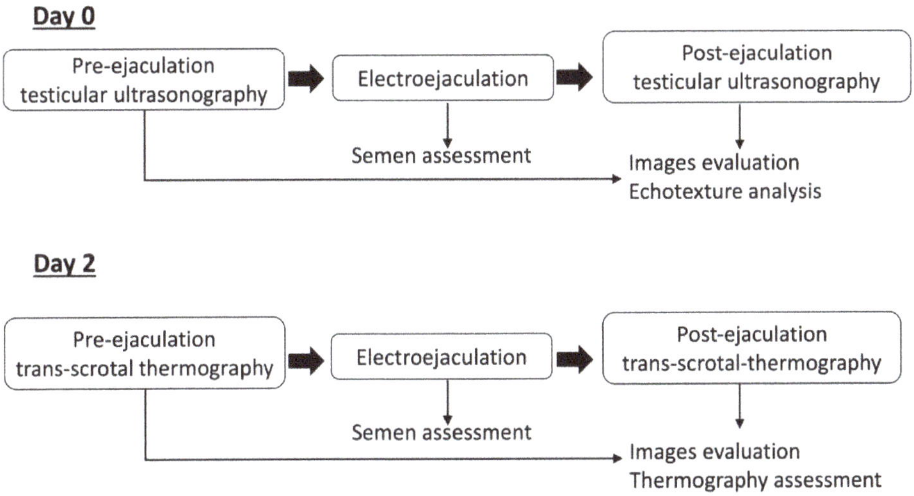

Figure 1. Scheme representation of the experimental design.

2.3. Testicular Examination and Scrotal Circumference Measurement

While restrained, each bull underwent testicular palpation to evaluate the testicular parenchyma and spermatic cords, with particular attention to testicular descent, consistency, size, and the presence of nodules or other abnormalities. Scrotal skin was also inspected for lesions or defects. Scrotal circumference was measured by encircling both testes at their widest point using a flexible metal tape with a measurement increment of 0.1 cm, covering a range from 1 to 60 cm (Figure 2). This measurement serves as a reliable indirect indicator of parenchyma testicular mass and total daily sperm production capacity.

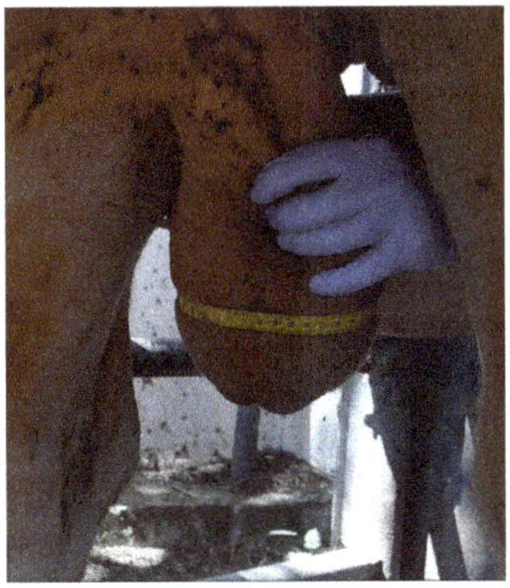

Figure 2. Flexible metal tape used for measuring the widest diameter of testes.

https://doi.org/10.3390/app16062780

2.4. Ultrasonographic Assessment

B-mode ultrasonography was performed using an Easy-Scan ultrasound unit (BCF Technology, Strathclyde, UK) equipped with a 3.0–7.0 MHz linear transducer. All examinations were conducted by the same operator under standardized settings (frequency: 7.0 MHz; gain: 8; depth: 12 cm).

On day 0, after restraining the bull, the transducer was positioned on each testis in both transverse and longitudinal planes. Six images were obtained per testis (12 images per bull): transverse scans at (1) the pampiniform plexus/head of the epididymis, (2) the proximal, (3) middle, and (4) distal portions of the testis, (5) the epididymal tail, and (6) one longitudinal scan of the testis. Ultrasonographic evaluations were performed both before and after semen collection.

2.5. Echotexture Analysis

Ultrasonographic images were analyzed using ImageJ software (version 1.6.0; U.S. National Institutes of Health, Bethesda, MD, USA). Two region-of-interest (ROI) selection methods were applied: (1) Single large ROI: The entire testis, including both the parenchyma and mediastinum, was selected to calculate the mean and standard deviation (SD) of pixel intensity (Figure 3). (2) Multiple small ROIs: Six circular ROIs (1 cm in diameter) were selected exclusively from the testicular parenchyma, and the mean ± SD of pixel intensity was calculated for each region (Figure 3). Pixel intensity values ranged from 0 (anechoic) to 255 (hyperechoic).

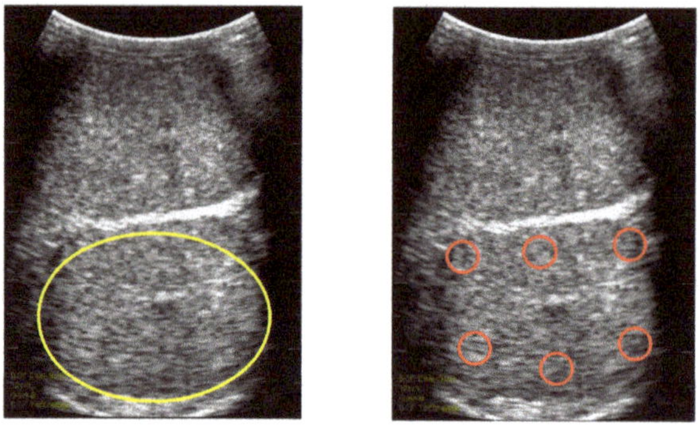

Figure 3. Ultrasound images showing both testes in transverse cross-section. On the (**left**), the whole-testis region of interest (in yellow) used for determining echotexture is shown. On the (**right**), the multiple-area method selecting six circular regions (in red) is illustrated.

2.6. Scrotal Thermography

Scrotal surface temperature was assessed using a thermal imaging camera (FLIR i7, FLIR Systems, Boston, MA, USA). Ambient temperature and relative humidity were recorded at the beginning and end of each session under stable environmental conditions, with values ranging from 10 to 16 °C and from 58% to 63%, respectively.

On day 2, thermographic images were acquired from a distance of 1 m behind the bull after a brief acclimation period in the chute to reduce stress. Images were analyzed using FLIR QuickReport software (version 1.2), and maximum, minimum, and mean temperatures were recorded for the entire scrotum as well as for the proximal and distal regions, both before and after ejaculation.

https://doi.org/10.3390/app16062780

2.7. Semen Collection and Evaluation

Bulls were restrained, preputial hair was trimmed, and the rectum was emptied prior to semen collection. Gentle, repeated rectal massage was performed to provide stimulation before EE. Semen was collected using an Electrojac 5.0 unit (Ideal Instruments®, Lexington, KY, USA).

The sperm-rich fraction was collected and evaluated macroscopically for volume, color, and density. Immediately after collection, a drop of semen was placed on a pre-warmed slide, and mass motility was subjectively assessed under a 40× objective using a 0–5 scale (0 = no movement; 5 = vigorous swirling) [29].

Semen samples were then diluted 1:2 with Bioxcell extender (IMV, L'Aigle, France), cooled to 5 °C, and transported to the laboratory. Sperm concentration was determined by diluting semen 1:100 in distilled water and counting spermatozoa using a Thoma chamber, with results confirmed by computer-assisted sperm analysis (CASA).

Sperm motility was evaluated using an automated image analysis system (ISAS v1; Proiser, Valencia, Spain). Samples were diluted to 25×10^6 sperm/mL, incubated at 37 °C for 10 min, loaded into ISAS D4C20 chambers, and analyzed under a 10× phase-contrast objective. Four randomly selected fields (approximately 200–300 spermatozoa per field) were recorded at 25 frames per second. Spermatozoa with head areas between 10 and 70 μm^2 were included in the analysis. Cells were classified as motile when the velocity average path (VAP) exceeded 10 μm/s and as progressively motile when straightness (STR) exceeded 75%. Percentages of total motility (TM) and progressive motility (PM) were recorded.

For sperm morphology assessment, smears were prepared using Hemacolor stain (Merck, Darmstadt, Germany) according to the manufacturer's instructions. A total of 100 spermatozoa per sample were examined under a 40× objective.

Based on semen quality parameters, bulls were classified as non-suitable (TM < 60%, PM < 30%, or normal morphology < 70%) or suitable (TM $\geq$ 60%, PM $\geq$ 30%, and normal morphology $\geq$ 70%).

2.8. Statistical Analysis

All data were analyzed using IBM SPSS Statistics software (version 29.0; IBM Corp., New York, NY, USA). Assumptions of linear mixed models (LMM) were evaluated by visual inspection of residual plots and Q–Q plots. No substantial deviations from normality or homoscedasticity were detected for the primary dependent variables. Descriptive statistics are presented as mean ± standard error (SE). Statistical significance was set at $p < 0.05$.

A LMM following a Restricted Maximum Likelihood (REML) approach was conducted to evaluate the effects of moment of evaluation in reference to the ejaculation (pre- and post-ejaculation), echotexture assessment methods (single whole-testis ROI vs. multiple parenchymal ROIs), probe situation in relation to the assessed testis (close and far), testicular cross-section (proximal, middle and distal), testes side (right and left) and age group (young and adult) on the mean echotexture values. Degrees of freedom were estimated using Satterthwaite's approximation. A similar statistical approach was used to evaluate the effect of evaluation moment (pre- and post-ejaculation), scrotal region (proximal and distal) and age group (young and adult) on the mean scrotal temperature measured by infrared thermography. Fixed effects were selected a priori based on biological relevance and to ensure model parsimony, given the limited sample size. To account for the hierarchical structure of the data and the correlation between repeated measurements taken from the same subject, each individual bull was included as a random intercept. Post hoc pairwise comparisons were conducted using Bonferroni adjustment.

https://doi.org/10.3390/app16062780

For variables that did not meet parametric assumptions, comparisons between independent groups, including young versus adult bulls and suitable versus non-suitable bulls, were conducted using the Mann–Whitney U test. Associations among echotexture parameters, thermographic measurements, scrotal circumference, and semen quality variables were assessed using Spearman's rank correlation coefficient (ϱ).

3. Results

3.1. Seminal Quality and Scrotal Circumference

Semen quality assessment indicated that five bulls (41.6%) did not meet the minimum criteria to be classified as suitable breeders. The primary deficiencies were low total motility, reduced progressive motility, or abnormal sperm morphology. Detailed seminal parameters and scrotal circumference (SC) values are presented in Table 1.

Table 1. Optimal ejaculate characteristics and scrotal circumference (SC) in Limousin bulls. Minimum values for semen quality characteristics: TM = 60%, PM = 30%, and normal morphology = 70%.

Id.	Age (Months)	SC (cm)	Volume (mL)	Sperm Concentration ($\times 10^6$ Spermatozoa/mL)	Total Motility (%)	Progressive Motility (%)	Normal Morphology (%)
1	14	30	4.5	432	60	43	76.3
2 *	14	33	8.0	704	12	8	70.4
3	14	32	7.0	656	61	42	86.3
4 *	13	34	5.5	320	40	22	79.2
5	60	36	15.0	1472	65	37	84.2
6	12	35	10.0	272	63	32	70.8
7 *	60	45	-	-	-	-	-
8 *	36	35	9.5	712	52	19	66.9
9	60	35	8.0	688	91	63	83.2
10 *	36	36	6.0	128	45	18	73.9
11	13	39	8.0	712	64	31	71.3
12	13	34	6.0	352	74	58	72.4

An asterisk (*) indicates bulls that did not meet the minimum quality criteria. (-) indicates azoospermia.

3.2. Testicular Ultrasonographic Findings in Testicles

Eleven bulls (91.7%) exhibited normal ultrasonographic patterns, characterized by a well-defined hyperechoic tunica albuginea and homogeneous, moderately echogenic parenchyma. The mediastinum testis appeared as a slightly hyperechoic linear structure measuring 2–4 mm in thickness. The epididymal head was visualized as a homogeneous, hypoechoic triangular region with hyperechoic margins. The epididymal body was less consistently visualized due to its medial location, whereas the epididymal tail appeared as a rounded, hypoechoic structure. The pampiniform plexus was located dorsolateral to the epididymal head and displayed a heterogeneous pattern composed of multiple tubular hypoechoic and anechoic structures.

A 5-year-old bull (Id. 7) presented pathological alterations, including a mildly enlarged left testis detected by palpation and bilateral fluid accumulation between the testis and scrotal sac, consistent with hydrocele. In addition, diffuse, small hyperechoic foci with acoustic shadowing—suggestive of testicular fibrosis or microlithiasis—were also observed. This bull was excluded from subsequent echotexture and thermographic analyses.

3.3. Testicular Echotexture

No significant differences in testicular echotexture were detected between bulls classified as suitable (142.4 ± 0.9) or non-suitable (138.6 ± 1.7) based on semen quality. Similarly, no significant differences were observed between the right and left testes (Table 2). However, in some bulls with poor semen quality, the echotexture difference between the right and left testis exceeded 20%, suggesting potential dysfunction in the testicular function (Figure 4).

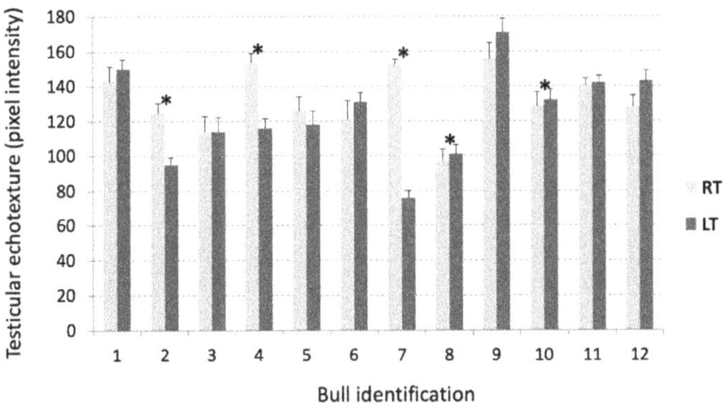

Figure 4. Average testicular echotexture (pixel intensity) in right (RT) and left (LT) testes in Limousin bulls. Data are expressed as mean ± SE. Asterisks indicate non-suitable bulls.

Table 2. Mean values (±SE) of testicular echotexture (pixel intensity) according to bull-related biological characteristics, including testicle side (right vs. left), and age group (young vs. adult).

	Testicle Side		Age Group	
	Right (n = 187)	Left (n = 192)	Young (n = 141)	Adult (n = 238)
	139.5 ± 13.6	141.9 ± 13.6	141.4 ± 12.6	145.0 ± 12.7
p value	n.s.		n.s.	

n.s. indicates non-significant differences.

Younger bulls (12–14 months) exhibited lower echotexture values compared to adult bulls (36–60 months), although these differences did not reach statistical significance ($p > 0.05$) (Table 2).

Testicular echotexture values increased significantly following semen collection, rising from mean values of 134.3 ± 13.6 before semen collection to 146.9 ± 13.6 after ejaculation ($p < 0.001$) (Table 3).

Table 3. Mean values (±SE) of testicular echotexture (pixel intensity) obtained under different methodological conditions, including the moment of evaluation (pre- and post-ejaculation), region of interest (single whole-testis area vs. multiple parenchymal areas), testicular cross-section (proximal, middle, distal), and distance between the probe and the testicle (close vs. far).

	Moment of Evaluation		ROIs		Testicular Cross-Section			Distance Probe-Testicle	
	Pre-Ejaculation (n = 186)	Post-Ejaculation (n = 193)	Single Area (n = 184)	Multiple Areas (n = 195)	Proximal (n = 126)	Middle (n = 126)	Distal (n = 126)	Close (n = 191)	Far (n = 188)
	134.3 ± 13.6	146.9 ± 13.6	138.4 ± 13.6	142.9 ± 13.6	142.0 ± 13.6	140.5 ± 13.6	139.4 ± 13.6	137.4 ± 13.6	143.9 ± 13.6
p value	***		**		n.s.			***	

n.s. indicates non-significant differences. Statistical significance is indicated as ** $p < 0.01$ and *** $p < 0.001$.

Echotexture values obtained using a single large ROI encompassing both the parenchyma and mediastinum were significantly lower (138.4 ± 13.6; $p < 0.01$) than those derived from six independent parenchymal ROIs (142.9 ± 13.6) (Table 3).

https://doi.org/10.3390/app16062780

No significant differences in echotexture were observed among proximal (142.0 ± 13.6), middle (140.5 ± 13.6) and distal (139.4 ± 13.6) testicular planes ($p > 0.05$) (Table 3).

When the testis closest to the transducer was evaluated, significantly lower pixel intensity values were recorded (137.4 ± 13.6; $p < 0.001$) compared with the testis located farther from the probe, using the contralateral testis as an acoustic window (143.9 ± 13.6) (Table 3).

No significant correlations were identified between testicular echotexture parameters and semen quality traits or SC.

3.4. Trans-Scrotal Thermography

Mean trans-scrotal temperature did not differ significantly between suitable and non-suitable bulls based on semen quality ($p > 0.05$). Scrotal temperature decreased significantly following ejaculation, dropping from 27.9 ± 2.6 °C pre-collection to 24.6 ± 2.6 °C post-collection (Table 4). A consistent temperature gradient of approximately 4 °C was observed between the proximal and distal regions of the scrotum (Table 4); the proximal region was consistently warmer than the distal region. Regarding age group, adult bulls exhibited significantly higher scrotal temperatures than young bulls ($p < 0.05$) (Table 4).

Table 4. Mean values (±SE) of temperature (°C) obtained under different methodological conditions, including the moment of evaluation (pre- and post-ejaculation), portion of testis (proximal and distal), and age group (young and adult).

	Moment of Evaluation		**Portion of Testis**		**Age Group**	
	Pre-Ejaculation (n = 22)	**Post-Ejaculation (n = 22)**	**Proximal (n = 22)**	**Distal (n = 22)**	**Young (n = 28)**	**Adult (n = 16)**
	27.9 ± 2.6	24.6 ± 2.6	28.1 ± 2.6	24.4 ± 2.6	25.6 ± 2.6	27.0 ± 2.6
p value	***		***		*	

Statistical significance is indicated as follows: * $p < 0.05$; *** $p < 0.001$.

No significant correlations were detected between scrotal thermographic variables and semen quality parameters or SC (Table 5).

Table 5. Spearman's rank correlation coefficients (ϱ), ranging from −1 (strong negative correlation) to +1 (strong positive correlation). Testicular echotexture and scrotal thermography showed no significant correlations with sperm quality parameters.

	Scrotal Circumference	**Total Sperm Motility**	**Progressive Sperm Motility**	**Sperm Morphology**
Testicular echotexture	−0.01 n.s.	−0.01 n.s.	0.01 n.s.	−0.01 n.s.
Scrotal thermography	0.09 n.s.	−0.07 n.s.	0.02 n.s.	−0.04 n.s.

n.s., not significant.

4. Discussion

In the present study, testicular ultrasonography was performed before and after semen collection to evaluate testicular parenchymal characteristics and to identify subclinical abnormalities that may not be detectable by routine physical examination. The results also allowed for the assessment of changes in testicular echotexture associated with ejaculation, providing insights into short-term functional responses of testicular tissue.

It has been estimated that 20–40% of bulls may be subfertile [30]. Accordingly, a relatively high incidence of testicular abnormalities was anticipated. However, among the twelve bulls included in this study (seven young and five adult), only one adult bull (8.3%) exhibited a detectable ultrasonographic lesion, whereas 42% showed low semen quality in the absence of clinical or ultrasonographic abnormalities. These findings suggest that subfertility in bulls is frequently functional rather than structural in nature. Given the limited sample size, these results should be interpreted with caution. The fibrotic lesion identified in one bull, characterized by a hyperechoic area with acoustic shadowing and disruption of seminiferous architecture, is consistent with previous reports [31]. Such lesions may originate from developmental disorders, infectious processes, thermoregulatory failure, or trauma, and their impact on semen quality largely depends on lesion extent and etiology [32,33].

Ultrasonography remains a cornerstone diagnostic tool for the detection of testicular abnormalities, including lesions not evident on palpation [34]. Beyond qualitative evaluation, quantitative analysis of pixel intensity (echotexture) provides objective information on testicular microstructure [30]. Although emerging approaches, such as artificial intelligence-based image analysis, show promise for improving diagnostic accuracy, methodological standardization remains essential to ensure reproducibility and clinical applicability [33].

Consistent with previous studies, the present findings indicate that testicular echotexture alone is not a reliable predictor of semen quality [24,35]. No associations were observed between echotexture parameters and spermatogenic indicators, including seminiferous epithelial cell populations, spermatogenic efficiency, or Sertoli cell indices [36]. Nevertheless, novel analytical tools, such as ECOTEXT software, may enhance predictive capacity once sufficiently validated [37,38].

No significant differences in echotexture were detected between the right and left testes in bulls classified as suitable breeders, in agreement with earlier reports [13,24,39]. However, marked echotexture asymmetry was observed in some bulls with poor semen quality, suggesting that individual testis evaluation may provide additional insights into reproductive potential. This finding supports the importance of bilateral assessment rather than reliance on averaged values.

The age-related differences observed in this study—specifically, lower echotexture values in younger bulls—are consistent with testicular maturation and increased development of seminiferous tubules with age [33,40,41]. Nonetheless, other studies have reported no association between age and echotexture [38,42], underscoring the influence of breed, management system, and environmental conditions on testicular echogenic patterns.

Several methodological factors were shown to influence echotexture measurements. Echogenicity values derived from were higher than those obtained from a single large ROI that included the mediastinum, supporting the recommendation to analyze multiple small, homogeneous regions for a more reliable representation of testicular tissue. If a single large ROI is used, strict consistency in ROI placement and tissue inclusion criteria is required to ensure comparability across examinations and studies.

With respect to scan plane selection, the present results indicate that echotexture measurements can be obtained from any testicular section in clinically normal testes. The middle testicular plane may be preferable in practice due to ease of acquisition and consistency. Higher echotexture values were recorded when the testis located farther from the transducer was evaluated, possibly due to acoustic attenuation or beam incidence effects. From a practical standpoint, assessing the testis closest to the probe may facilitate more consistent image acquisition.

A significant increase in echogenicity following ejaculation was observed. From a physiological perspective, this finding may reflect a transient response associated with fluid release from seminiferous ducts and temporary rearrangement of testicular architecture. In rams, echogenic variations have been linked to changes in seminiferous tubule lumen size [13] and to cellular differentiation within the seminiferous epithelium, although comparable evidence in bulls remains limited [43]. Because no histological evaluations were

https://doi.org/10.3390/app16062780

performed in the present study, the structural basis of these echotextural changes cannot be confirmed. Therefore, post-ejaculation changes in echotexture should be interpreted as short-term functional responses rather than permanent tissue alterations.

The mean scrotal surface temperature recorded in this study (25.9 °C) was slightly lower than values reported in some previous studies [42], although it was comparable to those in other studies conducted under similar environmental conditions [23,44]. The consistent temperature gradient of approximately 4 °C between the proximal and distal scrotal regions aligns with established thermoregulatory patterns [25,32,45].

No significant correlations were detected between scrotal surface temperature and semen quality, which is in agreement with several previous reports [3,24], although contrasting findings have been described elsewhere [46,47]. In dogs, a reduction in scrotal temperature following ejaculation has been attributed to testicular capsule contractions and transient reductions in testicular vascularity [48,49]. The approximately 3 °C decrease observed after ejaculation in the present study may reflect similar physiological mechanisms, including temporary testicular retraction and altered scrotal thermodynamics. Environmental conditions also exert a substantial influence on thermographic measurements; under cooler conditions, the distal scrotal pole dissipates heat more efficiently, whereas this gradient diminishes under warmer environments. Accordingly, accurate recording of ambient conditions is essential for an appropriate interpretation of thermographic data.

Traditional bull breeding soundness evaluations rely primarily on physical examination and semen analysis. Although still underutilized, echotexture and thermography offer non-invasive, objective, and complementary tools for assessing testicular structure and function. With continued advancements in imaging technology and data analysis, these methods may enhance diagnostic precision and support more informed selection decisions in breeding programs.

This study demonstrated that testicular echotexture and scrotal surface temperature in beef bulls undergo significant short-term changes following semen collection by electroejaculation. Specifically, testicular echogenicity increased, whereas scrotal surface temperature decreased by approximately 3 °C after ejaculation. However, neither ultrasonographic echotexture nor thermographic measurements were associated with semen quality parameters or scrotal circumference. These findings indicate that, although testicular ultrasonography and scrotal thermography cannot replace semen analysis for fertility assessment, they are capable of detecting transient functional changes and identifying subclinical or structural testicular alterations that may not be evident during routine clinical examination. Therefore, both techniques provide valuable complementary information within a comprehensive bull breeding soundness evaluation. Further studies involving larger populations and controlled environmental conditions are warranted to determine whether asymmetries in echotexture or thermographic patterns between testes may serve as early indicators of subfertility in bulls.

Author Contributions: Conceptualization, C.C.P.-M.; methodology, C.C.P.-M. and L.Q.-S.; investigation, C.C.P.-M. and L.Q.-S.; writing—original draft preparation, C.C.P.-M. and L.Q.-S.; writing—review and editing, C.C.P.-M. and L.Q.-S.; project administration, C.C.P.-M. All authors have read and agreed to the published version of the manuscript.

Funding: This research was supported by the Department of Animal Medicine and Surgery of the University of Cordoba, Spain. No external funding was received for this study.

Institutional Review Board Statement: The study was conducted in accordance with the guidelines set out in European Union legislation on animal experimentation (Directive 2010/63/EU) and transposed into Spanish law (RD 53/2013). The experimental protocol was reviewed and approved

https://doi.org/10.3390/app16062780

on July 6th, 2022, by the Animal Experimentation Ethics Committee of the University of Córdoba (approval no. 2022PI/20).

Informed Consent Statement: Not applicable.

Data Availability Statement: The corresponding author will provide the datasets used and analyzed during the current work upon reasonable request.

Conflicts of Interest: The authors declare no conflicts of interest.

References

1. Decuadro-Hansen, G. Use of testicular ultrasonography in the evaluation of the potential fertility of bulls. In Proceedings of the Maskana, 1st International Congress of Specialized Beef Cattle Production, Cuenca, Ecuador, 3–5 June 2015; pp. 9–19.
2. Draaisma, H.S. Application of Infrared Scrotal Thermography Under Field Conditions in Bulls Extensively Managed in Tropical Costa Rica, and Its Relationship with Spermiogramme, Clinical Variables and Final Breeding Soundness Classification. Master's Thesis, Utrecht University, Utrecht, The Netherlands, 2015.
3. Gil, A.R. Correlation Between Scrotal Thermography and Seminal Quality in Bulls from the Tropic Region of Central Veracruz State. Bachelor's Thesis, Universidad Veracruzana, Veracruz, Mexico, 2009.
4. Abdel-Razek, A.K.; Ali, A. Developmental changes of bull (Bos taurus) genitalia as evaluated by caliper and ultrasonography. *Reprod. Domest. Anim.* **2005**, *40*, 23–27. [CrossRef] [PubMed]
5. Gnemmi, G.; Lefebvre, R.C. Bull anatomy and ultrasonography of the reproductive tract. In *Practical Atlas of Ruminant and Camelid Reproductive Ultrasonography*; DesCôteaux, L., Gnemmi, G., Colloton, J., Eds.; Blackwell Publishing: London, UK, 2010; pp. 143–162.
6. Arteaga, A.A.; Barth, A.D.; Brito, L.F.C. Relationship between semen quality and pixel-intensity of testicular ultrasonograms after scrotal insulation in beef bulls. *Theriogenology* **2004**, *64*, 408–415. [CrossRef] [PubMed]
7. Ribadu, A.Y.; Nakao, T. Bovine reproductive ultrasonography: A review. *J. Reprod. Dev.* **1999**, *45*, 13–28. [CrossRef]
8. Aravindakshan, J.P.; Honaramooz, A.; Bartlewski, P.M.; Beard, A.P.; Pierson, R.A.; Rawlings, N.C. Pattern of gonadotropin secretion and ultrasonographic evaluation of developmental changes in the testis of early and late maturing bull calves. *Theriogenolgy* **2000**, *54*, 339–354. [CrossRef]
9. Ahmad, N.; Noakes, D.E.; Subandrio, A.L. B-mode real time ultrasonographic imaging of the testis and epididymis of sheep and goats. *Vet. Rec.* **1991**, *128*, 491–496. [CrossRef]
10. Cotticelli, A.; Navas, L.; Calabria, A.; Bifulco, G.; Campanile, G.; Peric, T.; Prandi, A.; D'Occhio, M.J.; Russo, M. Incorporation of testicular ultrasonography and hair steroid concentrations in bull breeding soundness evaluation. *Vet. Sci.* **2023**, *25*, 373. [CrossRef]
11. Bracco, C.; Gloria, A.; Contri, A. Ultrasound-based technologies for the evaluation of testicles in the dog: Keystones and breakthroughs. *Vet. Sci.* **2023**, *10*, 683. [CrossRef]
12. Pozor, M.; Morrissey, H.; Albanese, V.; Khouzam, N.; Deriberprey, A.; Macpherson, M.L.; Kelleman, A.A. Relationship between echotextural and histomorphometric characteristics of stallion testes. *Theriogenology* **2017**, *99*, 134–145. [CrossRef]
13. Montes-Garrido, R.; Riesco, M.F.; Anel-Lopez, L.; Neila-Montero, M.; Palacin-Martinez, C.; Boixo, J.C.; de Paz, P.; Ortega-Ferrusola, C.; Hassan, M.A.A.; Anel, L.; et al. Application of ultrasound technique to evaluate the testicular function and its correlation to the sperm quality after different collection frequency in rams. *Front. Vet. Sci.* **2022**, *9*, 1035036. [CrossRef]
14. Asadi, B.; Cuenca-Zaldívar, J.N.; Carcasona-Otal, A.; Herrero, P.; Lapuente-Hernández, D. Improving the reliability of muscle tissue characterization post-stroke: A secondary statistical analysis of echotexture features. *J. Clin. Med.* **2025**, *14*, 2902. [CrossRef]
15. Roberto, J.V.B.; Souza, B.B. Use of infrared thermography in veterinary medicine and animal production. *J. Anim. Behav. Biometeorol.* **2014**, *2*, 73–84. [CrossRef]
16. Gianesella, M.; Arfuso, F.; Fiore, E.; Giambelluca, S.; Giudice, E.; Armato, L.; Piccione, G. Infrared thermography as a rapid and non-invasive diagnostic tool to detect inflammatory foot diseases in dairy cows. *Pol. J. Vet. Sci.* **2018**, *21*, 299–305. [CrossRef] [PubMed]
17. Cai, Z.; Cui, J.; Yuan, H.; Cheng, M. Application and research progress of infrared thermography in temperature measurement of livestock and poultry animals: A review. *Comput. Electron. Agric.* **2023**, *205*, 107586. [CrossRef]
18. Ziproudina, N.; Ming, Z.; Hänninen, O.O.P. Plantar infrared thermography measurements and low back pain intensity. *J. Manip. Physiol. Ther.* **2006**, *29*, 219–223. [CrossRef]
19. McManus, C.; Tanure, C.B.; Peripolli, V.; Seixas, L.; Fischer, V.; Gabbi, A.M.; Menegassi, S.R.O.; Stumpf, M.T.; Kolling, G.J.; Dias, E.; et al. Infrared thermography in ani- mal production: An overview. *Comput. Electron. Agric.* **2016**, *123*, 10–16. [CrossRef]
20. Yin, M.; Ma, R.; Luo, H.; Li, J.; Zhao, Q.; Zhang, M. Non-contact sensing technology enables precision livestock farming in smart farms. *Comput. Electron. Agric.* **2023**, *212*, 108171. [CrossRef]

https://doi.org/10.3390/app16062780

21. Zhao, K.; Ge, S.; Chen, Y.; Li, Q.; Guo, M.; Nian, Y.; Ren, W. Non-contact temperature monitoring in dairy cattle via thermal infrared imaging and environmental parameters. *Agriculture* **2026**, *16*, 306. [CrossRef]
22. Purohit, R.C.; Hudson, R.S.; Riddell, M.G.; Carson, R.L.; Wolfe, D.F.; Walker, D.F. Thermography of the bovine scrotum. *Am. J. Vet. Res.* **1985**, *46*, 2388–2392. [CrossRef]
23. Kastelic, J.P.; Coulter, G.H.; Cook, R.B.; Wallins, G.L.; Entz, T. Environmental factors affecting measurement of bovine scrotal surface temperature with infrared thermography. *Anim. Reprod. Sci.* **1996**, *41*, 153–159. [CrossRef]
24. de Oliveira, A.P.L.; Andrade, K.A.; Uscategui, R.A.R.; Salcedo, Y.T.G.; Iared, F.; Melo, T.V.; Siqueira, J.B. Echotexture and testicular thermography and their correlation with the andrological examination in young Nelore bulls. *Proc. Indian Natl. Sci. Acad.* **2025**, *91*, 457–470. [CrossRef]
25. Kastelic, J.P.; Cook, R.B.; Coulter, G.H. Scrotal surface, subcutaneous, intratesticular and intraepididymal temperatures in bulls. *Theriogenology* **1995**, *44*, 147–152. [CrossRef]
26. Brito, L.F.C.; Silva, A.E.D.F.; Barbosa, R.T.; Kastelic, J.P. Testicular thermoregulation in Bos indicus, crossbred and Bos taurus bulls: Relationship with scrotal, testicular vascular cone and testicular morphology, and effects on semen quality and sperm production. *Theriogenology* **2004**, *61*, 511–528. [CrossRef] [PubMed]
27. Menegassi, S.R.O.; Pereira, G.R.; Dias, E.A.; Rocha, M.K.; Carvalho, H.R.; Koetz, C., Jr.; Oberst, E.R.; Barcellos, J.O.J. Infrared thermography as a noninvasive method to assess scrotal insulation on sperm production in beef bulls. *Andrologia* **2018**, *50*, e12904. [CrossRef]
28. Morrell, J.M. Heat stress and bull fertility. *Theriogenology* **2020**, *153*, 62–67. [CrossRef] [PubMed]
29. Nath, R. Cryopreservation of Buffalo Semen. Master's Thesis, Department of Animal Reproduction, Gynaecology and Obstetrics, Govind Ballabh Pant University of Agricultural and Technology, Pantnagar, India, 1988.
30. Kastelic, J.P.; Brito, L.F.C. Ultrasonography for monitoring reproductive function in the bull. *Reprod. Domest. Anim.* **2012**, *47*, 368–373. [CrossRef]
31. Barth, A.; Alisio, L.; Avilés, M.; Arteaga, A.; Campbell, J.; Hendrick, S. Fibrotic lesions in the testis of bulls and relationship to semen quality. *Anim. Reprod. Sci.* **2008**, *106*, 274–288. [CrossRef]
32. Brito, L.F.C.; Barth, A.D.; Wilde, R.E.; Kastelic, J.P. Testicular ultrasonogram pixel intensity during sexual development and its relationship with semen quality, sperm production, and quantitative testicular histology in beef bulls. *Theriogenology* **2012**, *78*, 69–76. [CrossRef]
33. Tomlinson, M.; Jennings, A.; Macrae, A.; Truyers, I. The value of trans-scrotal ultrasonography at bull breeding soundness evaluation (BBSE): The relationship between testicular parenchymal pixel intensity and semen quality. *Theriogenology* **2017**, *89*, 169–177. [CrossRef]
34. Gnemmi, G.; Lefebvre, R.C. Ultrasound imaging of the bull reproductive tract: An important field of expertise for veterinarians. *Vet. Clin. N. Am. Food Anim. Pract.* **2009**, *25*, 767–779. [CrossRef]
35. Pinho, R.O.; Costa, D.S.; Siqueira, J.B.; Martins, L.F.; Teixeira, L.A.; Guimarães, S.E.F.; Guimarães, J.D. Lack of relationship between testicular echotexture and breeding soundness evaluation in adult Nelore bulls. *Livest. Sci.* **2013**, *157*, 40–46. [CrossRef]
36. Dias, W.; Faria, F.J.C.; Fernandes, C.A.C.; Sampaio, B.F.B.; Santos, M.D.; Rodriguez, W.B.; Borges Silva, J.C.; Costa, D.S. Testicular echotexture is not a viable method to indirectly evaluate the spermatogenic parameters in Nelore bulls. *Austral J. Vet. Sci.* **2017**, *1*, 45–51. [CrossRef]
37. Carvajal-Serna, M.; Miguel-Jiménez, S.; Pérez-Pe, R.; Casao, A. Testicular ultrasound analysis as a predictive tool of ram sperm quality. *Biology* **2022**, *11*, 261. [CrossRef]
38. Marcantonio, S.; Echegaray, A.; Gnemmi, G.; Quiroga, H.; Capdevielle, E.F.; Alza, G.; Sara, R.; Miragaya, M. Echotexture of testicular parenchyma as a predictor of seminal quality in AI beef bulls. *Large Anim. Rev.* **2022**, *28*, 165–171.
39. Cardilli, D.J.; Toniollo, G.H.; Pastore, A.A.; Canola, J.C.; Oliveira, J.A.; Mercadante, M.E.Z. Testicular ultrasonography in young Nelore bulls raised in an extensive management system. *Arq. Bras. Med. Vet. Zootec.* **2012**, *64*, 75–82. [CrossRef]
40. Chandolia, R.K.; Honaramooz, A.; Omeke, B.C.; Pierson, R.; Beard, A.P.; Rawlings, N.C. Assessment of development of the testis and accessory glands by ultrasonography in bull calves and associated endocrine changes. *Theriogenology* **1997**, *48*, 119–132. [CrossRef] [PubMed]
41. Santana, A.; Pérez, J.; Fernandes, L.; Alberto, L.; Henry, M.; Chacón, L. Biometric and ultrasonographic aspects of testicular development in Guzerat (Bos taurus indicus) bulls. *Rev. Cient. Anim.* **2012**, *5*, 51–62.
42. Brito, L.F.C.; Silva, A.E.D.F.; Rodrigues, L.H.; Vieira, F.V.; Deragon, L.A.G.; Kastelic, J.P. Effect of age and genetics on characteristics of the scrotum, testes and testicular vascular cones, and on sperm production and semen quality in AI bulls in Brazil. *Theriogenology* **2002**, *58*, 1175–1186. [CrossRef]
43. Giffin, J.L.; Bartlewski, P.M.; Hahnel, A.C. Correlations among ultrasonographic and microscopic characteristics of prepubescent ram lamb testes. *Exp. Biol. Med.* **2014**, *239*, 1606–1618. [CrossRef]
44. Berry, D.P.; Evans, R.D.; Parland, S.M.C. Evaluation of bull fertility in dairy and beef cattle using cow field data. *Theriogenology* **2011**, *75*, 172–181. [CrossRef]

https://doi.org/10.3390/app16062780

45. Gábor, G.; Sasser, R.G.; Kastelic, J.P.; Coulter, G.H.; Falkay, G.Y.; Mézes, M.; Bozó, S.; Völgyi-Csik, J. Morphologic, endocrine and thermographic measurements of testicles in comparison with semen characteristics in mature Holstein-Friesian breeding bulls. *Anim. Reprod. Sci.* **1998**, *51*, 215–224. [CrossRef]
46. Coulter, G.H.; Senger, P.L.; Bailey, D.R.C. Relationship of scrotal surface temperature as measured by infrared thermography to subcutaneous and deep testicular temperature in the ram. *J. Reprod. Fertil.* **1988**, *84*, 417–423. [CrossRef]
47. de Ruediger, F.R.; Chacur, M.G.M.; Alves, F.C.P.E.; Oba, E.; De Amorim Ramos, A. Digital infrared thermography of the scrotum, semen quality, serum testosterone levels in Nelore bulls (Bos taurus indicus) and their correlation with climatic factors. *Semin. Cienc. Agrar.* **2016**, *37*, 221–232.
48. Tekin, K.; İnanç, M.E.; Özen, D.; Cil, B.; Olğaç, K.T.; Yılmaz, B.; Taşdemir, U.; Tuncer, P.B.; Büyükleblebici, S.; Daşkın, A.; et al. Use of infrared thermography during ejaculation process and its link with semen quality and freezability in dogs. *Animals* **2021**, *11*, 3023. [CrossRef]
49. Shafik, A.; Shafik, A.A.; Shafik, I.A.; El Sibai, O. Physiological considerations of the morphologic changes of the testicles during erection and ejaculation: A canine study. *Urol. Int.* **2007**, *79*, 262–266. [CrossRef]

https://doi.org/10.3390/app16062780

applied sciences

MDPI

Article

Palmaris Longus Muscle and Its Variations: Ultrasound, Anatomical and Histological Study with Clinical and Surgical Applications

Maribel Miguel-Pérez [1,*,†], Angel Muniz Delgado [1,†], Sara Ortiz-Miguel [2,3,4], Sandra Lopez-Prades [5,6], Juan Carlos Ortiz-Sagristà [7], Cristina Miralles [5], Núria Ruiz-Roig [8], Ingrid Möller [1], Carlo Martinoli [9] and Albert Pérez-Bellmunt [4,10]

1 Unit of Human Anatomy and Embryology, Department of Pathology and Experimental Therapeutics, Faculty of Medicine and Health Sciences (Bellvitge Campus), University of Barcelona, 08907 Barcelona, Spain; anmudel@gmail.com (A.M.D.)
2 Unit of Human Anatomy and Embryology, Department of Surgery and Medical-Surgical Specialities, Faculty of Medicine and Health Sciences (Clinic Campus), University of Barcelona, 08007 Barcelona, Spain
3 Physiotherapy Barcelona UB-UdG-Euses, Campus Health sciences of Bellvitge UB c/ Feixa LLarga s/n, 08907 l'Hospitalet de Llobregat, Spain
4 Actium Functional Anatomy Research Group, Bellaterra, 08193 Cerdanyola del Vallès, Spain
5 Pathology Department, Hospital Clinic Barcelona, 08036 Barcelona, Spain
6 August Pi i Sunyer Biomedical Research Institute (IDIBAPS), 08036 Barcelona, Spain
7 Anesthesia Department, Fundació Puigvert, 08025 Barcelona, Spain
8 Department of Pathology, University Hospital of Bellvitge, 08907 Barcelona, Spain
9 Cattedra di Radiologia "R"-DICMI, Universita di Genova, 16126 Genoa, Italy
10 Unit of Human Anatomy and Embryology, Department of Morphological Sciences, Faculty of Medicine, Universitat Autònoma de Barcelona, Bellaterra, 08193 Cerdanyola del Vallès, Spain
* Correspondence: mimiguel@ub.edu
† These authors contributed equally to this work.

Abstract

Background/objective: The palmaris longus muscle is a variable and often inconsistent muscle in the anterior compartment of the forearm. This fusiform-shaped muscle originates at the medial epicondyle of the humerus bone following a long and narrow tendon that inserts at the palmar aponeurosis. That tendon is used in reconstructive surgery, and for this reason, detailed information from an ultrasound is used to detect the tendon and the possible variations in the muscle. The present study aimed to investigate the palmaris longus muscle and its variations through ultrasound, anatomical, and histological analysis with clinical and surgical applications. **Methods**: A total of 72 upper limbs from 33 females and 39 males, 32 right and 40 left, were evaluated in ultrasound, anatomical, and histological studies. The main objective was to prove the existence of the palmaris longus muscle and its variations, as well as to measure the tendon for surgical applications. **Results**: Ultrasound analysis showed that it is possible to determine the existence of the muscle (76.4%) and its variations (23.6%), as well as its absence (15.3%). The anatomical results proved the ultrasound results. The width of the tendon was between 0.4 and 0.38 mm. by ultrasound and anatomical analysis. Also, normal palmaris longus tendons were not a direct cause of compression of the median nerve. **Conclusions:** It is important to confirm the existence and possible variations in the palmaris longus muscle and tendon through ultrasound before surgical reconstruction and for clinical diagnostics.

Keywords: palmaris longus muscle variations; median nerve relationships; ultrasound

1. Introduction

The palmaris longus muscle (PL) is a variable and often inconsistent structure located in the most superficial layer of the anterior compartment of the forearm [1]. It originates as a small, fusiform, and slender muscle from the anterior and inferior surface of the medial epicondyle of the humerus bone (ME) [2,3], as well as from the adjacent intermuscular septum and the antebrachial fascia (AF) [4]. Distally, the muscle continues as a long tendon that lies medial to the tendon of the flexor carpi radialis (FCR) and the median nerve (MN) [5,6]. Near its insertion, the tendon gives off fibrous slips to the flexor retinaculum (FR); however, the main portion widens, passes superficially over the FR [7], and inserts into the palmar aponeurosis (PA). The PL is innervated by the MN [2,3], and its principal motor function is to assist in wrist flexion and, to a lesser extent, in thumb abduction [8]. In addition, due to its high tendinous content, a proprioceptive role has also been suggested [9].

This muscle is present only in mammals that employ the upper limb for locomotion. It may be absent in humans and great apes, though it is consistently present in the orangutan [5]. Its presence in humans varies widely (from 1.5% to 63%) depending on population and ethnicity, and agenesia is its most frequent anatomical variation [1,7,8]. Numerous morphological variants have been reported, including accessory, diminutive, or reversed bellies, as well as tendons with atypical, bifid, or multiple insertions [10]. These variants are not merely academic; certain PL variants have been implicated in the compression of adjacent neurovascular structures, such as the median or ulnar nerve or vessels, potentially producing entrapment syndromes at the level of the wrist [6,11]. Also, it can even mimic soft-tissue tumors. In response, recent studies have proposed new classification schemes to catalog PL morphology, reflecting a growing recognition that such variations must be understood in a clinical and surgical context [6].

The clinical importance of the PL lies in its extensive use in reconstructive surgery [12,13]. Because of its favorable length, diameter, and superficial position (and owing to its vestigial function), it is often selected as a donor tendon for grafting procedures [14], including the reconstruction of digital pulleys [15], ligaments, and tendons, as well as for the restoration of the lips and chin [1,10]. For these reasons, accurate differentiation of the PL tendon from adjacent structures, particularly the MN at the level of the carpal tunnel, is essential for safe harvesting [14,16]

Indeed, knowledge of the PL's anatomical variations is crucial for operative planning, as an anomalous or absent PL can influence the surgical approach and graft selection [16]. In this context, ultrasound (US) has emerged as a non-invasive, accessible, and reliable technique for identifying the PL and assessing its tendinous characteristics [14,15].

Histologically, the PL's muscle belly contains a heterogeneous mix of fiber types (type I and II) [17], a feature that underscores its functional adaptability. This mix of fatigue-resistant and fast-twitch fibers may facilitate the muscle's role in dynamic reconstructive surgeries, helping transferred tendons integrate and restore motion effectively [17]. In summary, appreciating the spectrum of PL variants (from gross anatomy to microscopic composition) is of paramount importance for clinicians utilizing this "spare" tendon in modern reconstructive surgery.

Therefore, the aim of this study is to provide comprehensive ultrasound, anatomical, and histological descriptions of the palmaris longus muscle, its tendon, and its possible variations, detailing its course from origin to insertion and its special relationships with the adjacent MN, in order to facilitate the precise identification of these structures and improve surgical planning.

https://doi.org/10.3390/app16020817

2. Materials and Methods

An ultrasound, anatomical, and histological study was performed on 72 upper limbs (40 left, 32 right; 33 females, 39 males; average age 80 years) from fresh, frozen, and thawed specimens cryopreserved at −20 °C in the dissection room. No scars were observed in any of the cadavers studied. All the specimens were obtained from donated corpses for anatomic research and teaching from the dissection room of the Faculty of Medicine and Health Sciences. These specimens were donated by the donor in life, signing a written informed consent form. Every effort was made to comply with local and international ethical guidelines, including the Declaration of Helsinki and laws concerning the use of human cadaveric donors in anatomical research. Approval was granted by the Local Ethics Committee (Institutional Review Board IRB00003099, signed 28 February 2022).

2.1. Ultrasound Study

The US examinations were performed using a LOGIQ P9 ultrasound system (GE Ultrasound Korea Ltd., Seongnam, Republic of Korea) equipped with a high-frequency linear transducer (6–15 MHz). Scans were obtained in both the transverse and longitudinal planes with the upper limb positioned anatomically and the forearm in full supination.

The origin of the flexor muscles at the ME was identified, and the PL was visualized when it was present. Variations were classified as follows: Type 0 (absence of the PL); type 1 (presence of a purely tendinous structure without a muscular belly); type 2 (presence of a muscular belly located in the middle third of the forearm, with both proximal and distal tendons); type 3 (presence of a distal muscular belly).

Additional anatomical variants associated with the PL were also recorded. The relationships between the PL and the surrounding musculature or deep fascia, as well as its relationship with the MN at the wrist, were evaluated. The width of the tendon of the PL was measured at two points: 3 cm from the origin of the muscular belly and 2 cm proximal to its insertion into the PA. At the wrist, the widths of both the PL tendon and the MN were recorded.

2.2. Anatomical Study

Subsequent anatomical dissections were performed following classical procedures. A longitudinal incision was made along the midline of the anterior forearm, with transverse incisions proximal to the elbow and distal to the wrist. The existence and possible variations in the PL were assessed first. When present, its relationships with the AF were examined. After removal of the fascia, morphometric measurements were obtained using a digital vernier caliper. All measurements were performed three times by the same examiner, and the mean of the two closest values was used for analysis.

The following parameters were recorded (Figure 1):

A. Length of the forearm (from the ME to the styloid process (SP) of the ulnar bone)

B. Length of all PL from the ME origin to the insertion at the PA.

C. Length of the muscular bell of the PL from the ME to the end of the muscular fibers.

D. Length of the tendon from the origin at the distal part of the muscular bell to the insertion at the PA.

E. Width of the muscle of PL at 3 cm from the ME.

F. Width of the muscle of PL at 6 cm from the ME.

G. Width of the tendon of PL at 3 cm from the origin (end of the belly muscle)

H. Width of the tendon of PL 2 cm before arrived at its insertion at the PA.

I. Distance from the SP of the radius bone to the lateral border of the tendon of PL

J. Distance from the styloid process of the radius bone to the medial border of MN

K. Width of the MN at the level of the wrist, before it is introduced at the carpal tunnel.

https://doi.org/10.3390/app16020817

Before its insertion into the PA, the fascial expansions of the tendon and its relationships with the MN were carefully documented. All dissections were photographed using a Canon EOS 60D digital camera (Canon, Tokyo, Japan).

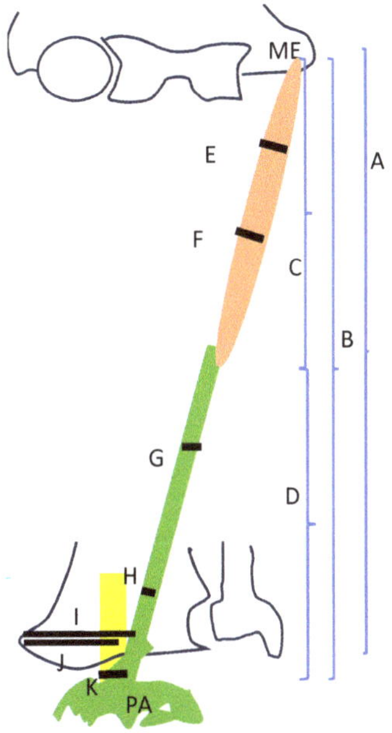

Figure 1. Diagram of the palmaris muscle and tendon with the different measures taken.

2.3. Statistical Study

All data were processed with Excel and SSPS 26. Qualitative variables were summarized as proportions. The concordance index (CCI) of the different measurements observed (nerve perimeter and diameter) was calculated for each anatomical point studied. If optimal CCI values were obtained, they were averaged and summarized as median and mean with their respective measures of dispersion.

2.4. Histological Study

In two randomly selected specimens, tissue examples of the PL tendon were obtained at multiple points. Both the longitudinal and transversal samples of the PL tendon were formalin-fixed and paraffin-embedded to obtain a paraffin block, from which 4 micras Haematoxilin and Eosin (H&E) slides were obtained, and a histological analysis was performed by a soft-tissue-specialized pathologist. The samples were analyzed using an optical microscope (Leica DMD 108, Leica, Wetzlar, Germany).

3. Results

A total of 72 upper limbs were studied. 39 males (54.2%) and 33 females (45.8%), with 40 from the left side (55.6%) and 32 from the right side (44.4%).

In total, 55 upper limbs presented PL (76.4%), and in 17 upper limbs, we observed anatomic variations in the PL (23.6%). The statistical study showed that the anatomical variations were distributed in the same way in males and females (Exact proof from the Fisher test = 0.408; $p = 0.237$) (Table 1) or sides (Exact prove de Fisher= 0.418; $p = 0.279$), although the male and left hands presented more variations (Table 2). The length of the forearm was 25.13 cm, SD 2.425 cm. The other parameters studied are shown in Table 3.

https://doi.org/10.3390/app16020817

Table 1. Variation in the palmaris longus muscle depending on sex.

		No Variation	Variation	Total
Sex	male	28	11	39
	female	27	6	33
Total		55	17	72

Table 2. Variation in the palmaris longus muscle depending on the side.

		No Variation	Variation	Total
Side	D	26	6	32
	I	29	11	40
Total		55	17	72

Table 3. Measures of the variable studied in the normal palmaris longus muscle.

Variable	Mean	SD	IC 95%	Median	IQR
A: Length of the forearm from the medial epicondyle to the ulnar styloid process	25.13	2.425	24.46 to 25.79	24.80	4
B: Length of all palmaris longus (muscle and tendon)	12.01	1.708	11.55 to12.48	12.04	2
C: Length of the muscular bell of the palmaris longus muscle	13.81	2.404	13.15 to 14.47	13.60	4
D: Length of the tendon of the palmaris longus muscle	25.82	2.394	25.17 to 26.48	25.58	4
E: Width of the palmaris longus muscle at 3 cm from the medial epicondyle	1.09	0.335	1.00 to 1.119	1.10	0
F: Width of the palmaris longus muscle at 6 cm from the medial epicondyle	1.04	0.315	0.9 to 1.13	1.08	1
Gu: Width of the tendon of palmaris longus at the muscle 3 cm from the origin measured by ultrasound	0.39	0.09	0.37 to 0.41	0.40	0.09
Ga: Width of the tendon of palmaris longus at the muscle 3 cm from the origin, measured by ultrasound and anatomy	0.38	0.065	0.36 to 0.40	0.38	0
Hu: Width of the tendon of palmaris longus at 2 cm before insertion at the palmar aponeurosis measured by ultrasound	0.39	0.079	0.37 to 0.41	0.39	0
Ha: Width of the tendon of palmaris longus at 2 cm before insertion at the palmar aponeurosis measured by anatomy	0.37	0.082	0.34 to 0.39	0.38	0
I: Distance from the styloid process of the radio bone to the lateral border of the palmaris longus tendon	2.35	0.305	2.27 to 2.44	2.34	0

https://doi.org/10.3390/app16020817

Table 3. *Cont.*

Variable	Mean	SD	IC 95%	Median	IQR
J: The styloid process of the radio bone to the medial border of the median nerve	2.54	0.315	2.46 to 2.63	2.50	1
Distance between the median nerve and the palmaris longus tendon	−0.13	0.472	−0.25 to 0.00	−0.11	0
Ku: Median nerve width measured by ultrasound	0.55	0.067	0.53 to 0.56	0.55	0
Ka: Median nerve width measured by anatomy	0.52	0.069	0.50 to 0.53	0.51	0

3.1. Ultrasound and Anatomical Study

Ultrasound study was able to identify the presence of PL in 55 forearms with 17 variations, ten being type 0, and seven other variations. Dissections proved the normal pattern visualized by US, the absence of the muscle, and the rest of the variations in the PL (Table 3).

Ultrasonography findings were consistent with the anatomical study in 98.61%. Only in one case was the US not able to detect a type 1 variation in the PL, which was classified as type 0 by US.

When the PL existed, a US study showed its proximal insertion at the medial epicondyle. However, it was necessary to move down slightly in the inferior direction at the medial epicondyle to observe as a small and triangular hypoechogenic structure, with a superior base and inferior vertex. It was always surrounded by a hyperechogenic line that separated it from the FCR laterally and FCU medially (Figure 2).

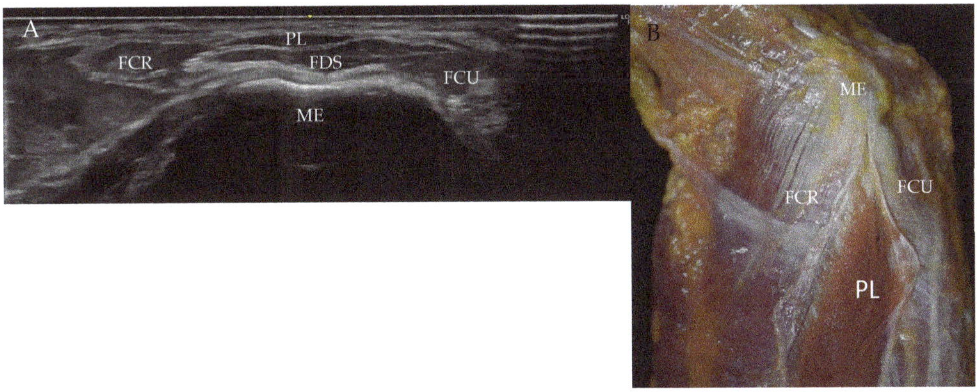

Figure 2. Origin of the palmaris longus muscle (PL) (**A**): Ultrasound image of the PL as a small and triangular hypoechogenic structure, with base superior and vertex inferior, surrounded by the hyperechogenic line (the antebrachial fascia) and separated from the rest of the muscles. Flexor carpi radialis (FCR), flexor carpi ulnaris (FCU), flexor digitorum superficialis (FDS), and medial epicondyle (ME). (**B**): Anatomical view of the origin of the PL. It is possible to observe how the muscular fibers do not arrive until the medial epicondyle.

After, the belly increased, and an ovoid shape was adopted, with a characteristic and transversal hyperechogenic line inside the muscle (Figure 3). This line became the tendon that followed the muscle distally. This tendon was a uniform and fibrillar hyperechogenic structure until it arrived and inserted at the PA.

https://doi.org/10.3390/app16020817

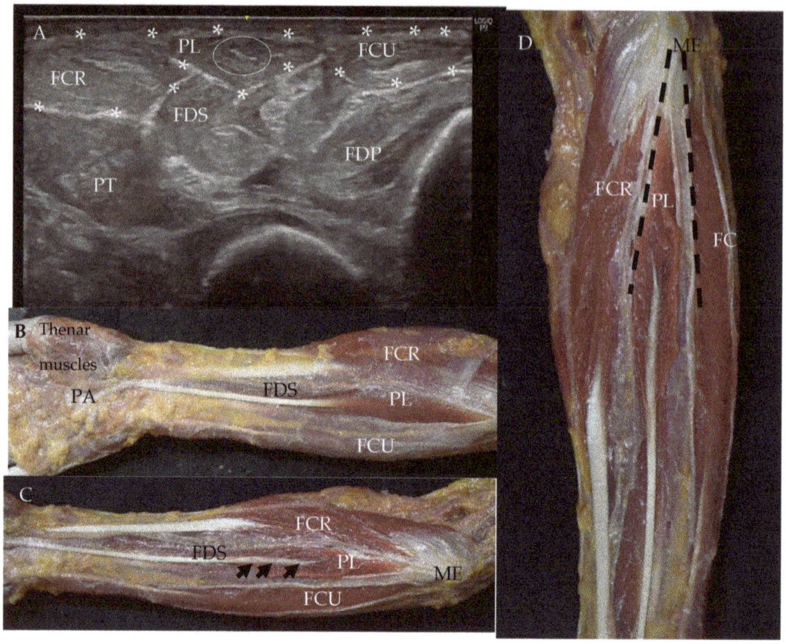

Figure 3. (**A**): Ultrasound vision of the normal palmaris longus muscle (PL) at the proximal third of the forearm. Antebrachial fascia (AF), seen as a hyperechogenic line surrounding (white *) the muscle and separating it from flexor carpi radialis (FCR), and flexor carpis ulnaris (FCU) also isolated by the AF. Inside the PL, a transverse hyperechogenic line is seen (white circle). It corresponds to the origin of the tendon of the PL. Flexor digitorum superficialis muscle (FDS), flexor digitorum profundus muscle (FDP), and pronator teris muscle (PT). (**B**,**C**): Anatomical vision of PLB: The AF has removed only over the PL. (**C**): The AF is cut just over the PL, FCRFCU. All of them originated at the medial epicondyle (ME). The muscular fibers of the PL were cut to see if its tendon was inside the muscle (black arrow). FDS Palmar aponeurosis (PA). (**D**): Anterior side of the forearm without the AF that covers FCR, PL, and FCU. The tendon of the PL is seen inside the muscle. The septum from the antebrachial fascia separates these muscles (black intermittent line).

Nevertheless, the tendon had a different relationship with the AF at the proximal part of the forearm than at the distal part. First, the tendon was surrounded by the AF, as were all the anterior muscles of the forearm, and at the third distal of the forearm, the tendon was isolated and surrounded totally by the AF, as were the tendons of the FCR and flexor carpi ulnaris (FCU) (Figure 4).

https://doi.org/10.3390/app16020817

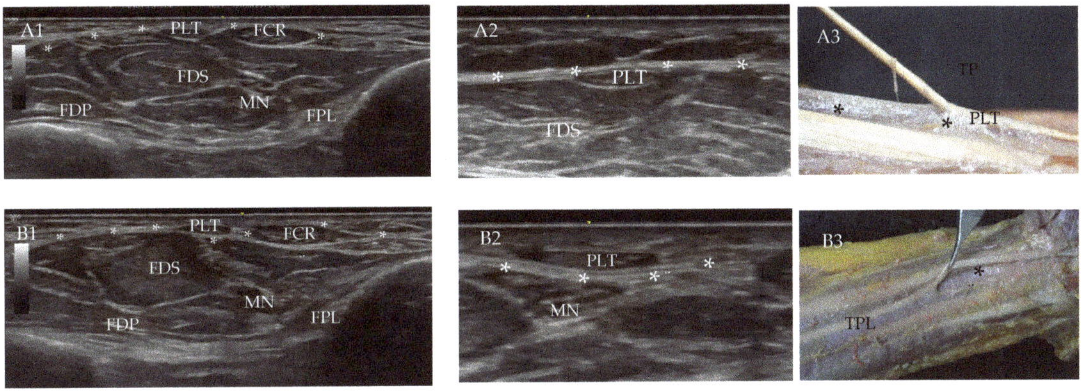

Figure 4. Relationship of the antebrachial fascia (AF) with the tendon of the palmaris longus muscle (PLT). (**A1**): Ultrasound (US) study shows that the AF (white *) covers the PTL and the rest of the anterior structures of the forearm. (**A2**): Enlarged image B, more specific. (**A3**): Anatomic image of the tendon of PL inside the AF (black *). (**B1**): US image of the tendon of the PL superficial and independent of the rest of the structures of the anterior side of the forearm. (**B2**): enlarge image. (**B3**): Anatomic view that shows the AF (black *) was under the tendon of PL at the distal third of the forearm surrounded the FDS, FDP, and MN.

Measurements of the width of the tendon at the two points studied were 0.39 SD 0.09 mm and 0.39 SD 0.079 mm. The width of the MN at the wrist was 0.55 SD 0.067 mm. There were no significant differences between the side of the body the limb was taken from or the sex (Table 4).

Table 4. Measures of Figure 1 of all the upper extremities that had the palmaris longus muscle, depending on the sex. Statistically significant values are in red.

Variable	Male	Female	Mean	IC 95%	T Student	p
A: Length of the forearm from the medial epicondyle to the ulnar styloid process	26.77	23.53	3.235	2.250 to 4.221	6.585	0.000
B: Length of all palmaris longus (muscle and tendon)	27.34	24.36	2.982	1.959 to 4.005	5.848	0.000
C: Length of the muscular bell of the palmaris longus muscle	12.48	11.57	0.901	0.008 to 1.795	2.023	0.048
D: Length of the tendon of the palmaris longus muscle	14.89	12.78	2.081	0.905 to 3.256	3.553	0.001
E: Width of the palmaris longus muscle at 3 cm from the medial epicondyle	1.10	1.07	0.039	−0.145 to 0.223	0.425	0.672
F: Width of the palmaris longus muscle at 6 cm from the medial epicondyle	1.02	1.06	−0.039	−0.212 to 0.135	−0.446	0.657
Gu: Width of the tendon of the palmaris longus at 3 cm from the origin measured by ultrasound	0.43	0.37	0.06	0.0246 to 0.0954	3.41	0.0014
Ga: Width of the tendon of the palmaris longus at 3 cm from the origin, measured by ultrasound and anatomy	0.41	0.35	0.059	0.027 to 0.091	3.688	0.01

https://doi.org/10.3390/app16020817

Table 4. *Cont.*

Variable	Male	Female	Mean	IC 95%	T Student	*p*
Hu: Width of the tendon of the palmaris longus at 2 cm before insertion at the palmar aponeurosis measured by ultrasound	0.40	038	0.023	−0.02 to 0.065	1.081	0.285
Ha: Width of the tendon of the palmaris longus at 2 cm before insertion at the palmar aponeurosis, measured by anatomy	0.37	0.36	0.010	−0.034 to 0.054	0.451	0.654
I: Distance from the styloid process of the radio bone to the lateral border of the palmaris longus tendon	2.50	2.23	0.269	0.116 to 0.422	3.522	0.001
J: the styloid process of the radio bone to the medial border of the median nerve	2.70	2.40	0.297	0.146 to 0.449	3.931	0.000
Distance between the median nerve and the palmaris longus tendon	−0.10	−0.15	0.046	−0.209 to 0.301	0.359	0.721
Ku: Median nerve width measured by ultrasound	0.53	0.56	−0.022	−0.058 to 0.014	−1.233	0.223
Ka: Median nerve width measured by anatomy	0.50	0.53	−0.034	−0.07 to 0.002	−1.886	0.065

In all the cases studied, the PL kept a great relationship with the MN at the wrist before the nerve was introduced inside the carpal tunnel, just up to the nerve or medially to it.

The anatomical study showed the normal pattern of PL in 55%.

The length of the muscular bell and tendon of PL was 25.82 cm, SD 2.394 cm (Table 3).

The muscle originated at the medial epicondyle, which has a thick aponeurosis where few muscular fibers originate, and at the AF (Figure 2). We did not find a real tendon at this point of origin. The muscular bell, with a fusiform shape, was located on the anterior and superficial sides of the forearm, between the FCR and FCU, covered by the AF (Figure 3B,C). Significant differences in the length of the muscle were found depending on the sex, but not on the side studied. Samples were bigger in males (Table 4).

The PL was separated by a septum of dense connective tissue coming from the AF (corresponding to the septum hyperechogenic seen by US) and the FCR and FCU (Figure 3D). Although the size and length of all the bell muscle was not consistent, there was always the same pattern. The muscular belly increases its width distally until the middle of the muscle, and then it decreases its width until the muscular fibers disappear, followed by a flat tendon. The length of the PL was 12.01 cm, SD 1.708 cm, the diameter of PL at the superior point was 1.09 cm, SD 0.335 cm, and the diameter at the inferior point was 1.04 cm, SD 0.315 cm. Significant differences in the length of the muscle were found depending on the sex (Table 4). The muscle was supplied by a branch of the MN that came deeply into the muscle (Figure 5).

An accurate dissection of the bell revealed the presence of an intramuscular tendon that corresponded to the hyperechogenic transversal line observed inside the muscle, as the US study showed. It was followed by the tendon of the PL (Figure 3C). Distally, the tendon runs superficially along the anterior side of the forearm, covered by the muscular fascia or AF. Nevertheless, as the US study exposed, this fascia had a different behavior in the middle than in the third distal part of the forearm, where the AF covered and isolated the tendon of the PL (Figures 4 and 6).

https://doi.org/10.3390/app16020817

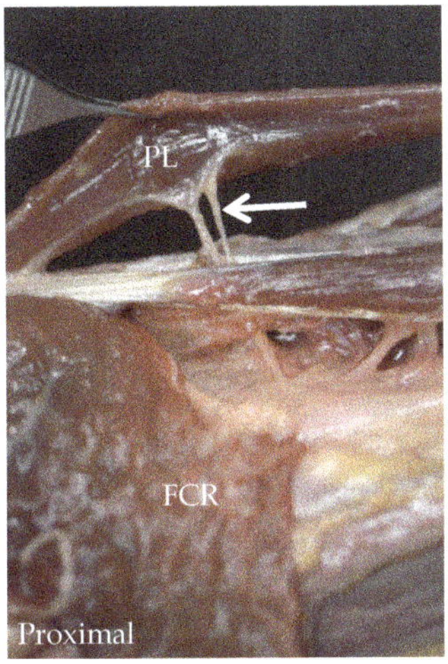

Figure 5. Anatomical view of the deep side of the palmaris longus muscle (PL) after removing the flexor carpi radialis muscle (FCR). Two branches of the median nerve (white arrow) cross the flexor digitorum superficialis muscle to supply the PL.

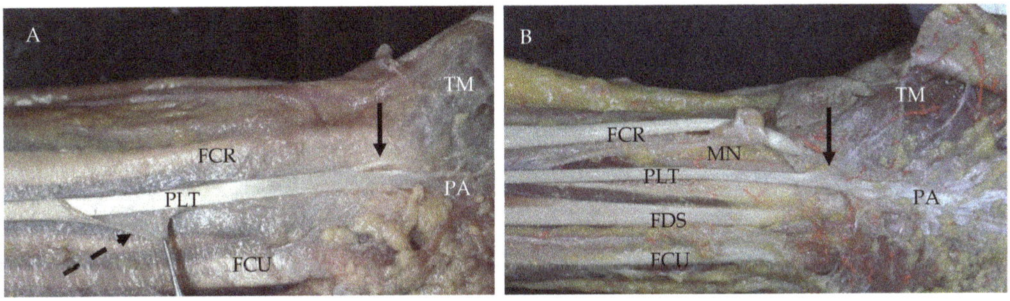

Figure 6. Anatomical dissection of the antebrachial fascia (AF) at the anterior side and distal third of the forearm (**A**): the AF (intermittent black arrow) cut only on the tendon of the palmaris longus muscle (PLT). It is possible to observe that it does not affect the rest of the muscles. However, under the PLT, the AF covers the rest of the anterior structures of the forearm. Thenar muscles (TM), palmar aponeurosis (PA). (**B**): All the AF has been removed. The PLT inserts at the PA and expansion to the TM is seen (black arrow). Flexor carpi radialis muscle (FCR), flexor carpi ulnaris muscle (FCU), median nerve (MN), flexor digitorum superficialis (FDS).

At this point, the tendon lost its deep relationship with the epimysium of FDS and was isolated from the rest of the muscles, nerves, and vessels of the anterior side of the forearm, as were the FCR and FCU. The AF made a special sheath for these three tendons until they arrived at the hand (Figure 3). Under these sheaths, the AF followed, covering the rest of the muscles, vessels, and nerves, and even increasing this thickness.

The length of the tendon was 13.81 cm, SD 2.404 cm (Table 3). Significant differences in the total length of the tendon were observed depending on the sex, but not on the side studied (Table 5).

https://doi.org/10.3390/app16020817

Table 5. Measures of Figure 1 of all the upper extremities that have the palmaris longus muscle, depending on the side. No statistically significant results were found.

Variable	Right	Left	Median	IC95%	T Student	p
A: Length of the forearm from the medial epicondyle to the ulnar styloid process	24.73	25.58	−0.844	−2.154 to 0.466	−1.293	0.202
B: Length of all palmaris longus (muscle and tendon)	25.47	26.24	−0.767	−2.064 to 0.530	−1.187	0.241
C: Length of the muscular bell of the palmaris longus muscle	11.98	12.08	−0.094	−1.022 to 0.834	−0.203	0.840
D: Length of the tendon of the palmaris longus muscle	13.49	14.16	−0.0673	−1.969 to 0.622	−1.042	0.302
E: Width of the palmaris longus muscle at 3 cm from the medial epicondyle	1.06	1.11	−0.043	−0.228 to 0.141	−0.469	0.641
F: Width of the palmaris longus muscle at 6 cm from the medial epicondyle	1.03	1.06	−0.031	−0.204 to 0.143	−0.357	0.723
Gu: Width of the tendon of the palmaris longus at 3 cm from the origin, measured by ultrasound	2.20	0.40	1.798	−1.602 to 5.198	1.061	0.294
Ga: Width of the tendon of palmaris longus at 3 cm from the origin, measured by ultrasound and anatomy	0.38	0.38	0.001	−0.035 to 0.037	0.043	0.966
Hu: Width of the tendon of the palmaris longus at 2 cm before insertion at the palmar aponeurosis, measured by ultrasound	0.39	0.40	−0.007	−0.05 to 0.036	−0.314	0.755
Ha: Width of the tendon of palmaris longus at 2 cm before insertion at the palmar aponeurosis, measured by anatomy	0.37	0.37	−0.001	−0.045 to 0.044	−0.035	0.972
I: Distance from the styloid process of the radio bone to the lateral border of the palmaris longus tendon	2.29	2.43	−0.133	−0.299 to 0.034	−1.601	0.115
J: the styloid process of the radio bone to the medial border of the median nerve	2.52	2.58	−0.058	−0.230 to 0.114	−0.675	0.502
Distance between the median nerve and the palmaris longus tendon	−0.20	−0.06	−0.144	−0.397 to 0.108	−1.148	0.254
Ku: Median nerve width, measured by ultrasound	0.55	0.54	0.011	−0.025 to 0.048	0.635	0.528
Ka: Median nerve width, measured by anatomy	0.52	0.51	0.005	−0.032 to 0.043	0.285	0.777

https://doi.org/10.3390/app16020817

The diameter of the tendon of the PL was 0.38 mm, SD 0.065 mm, at 3 cm from its origin and 0.37 mm, SD 0.082 mm, at 2 cm before the wrist line (Table 3). No significant differences in the diameter were found depending on the sex or side.

We found an inverse linear relationship between the length of the muscle and the diameter of the tendon of the PL (Spearman's rho = −0.349; $p = 0.009$). However, in a simple sampling simulation (number of samples = 1000), the statistical significance of the relationship would be lost (Spearman's rho = 0.245; $p = 0.071$) (Figure 7).

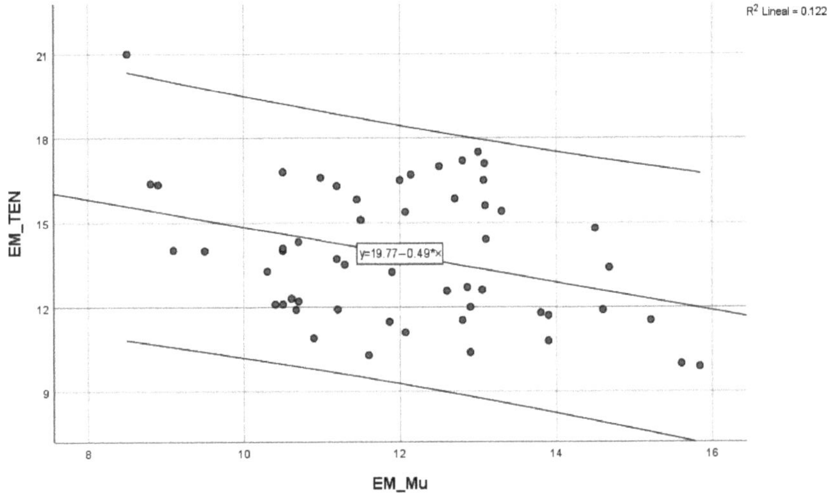

Figure 7. An inverse linear relationship exists between the length of the muscle and the tendon of the palmaris longus (Rho Spearman =−0.349; $p = 0.009$). However, in a simulation of the sample rate (n = 1000), the statistical significance of this relationship is lost (Rho Spearman = −0.245; $p = 0.071$).

The distance to the SP of the radio bone to the lateral border of the tendon of the PL at the level of the wrist was 2.35 cm, SD 0.305 cm, and to the medial border of the MN, it was 2.54 cm, SD 0.315 cm. Significant differences were observed depending on sex. Samples were bigger in males (Table 5). In 71% of the cases, the results were negative. This implies that the tendon of the PL covered the MN. There were significant differences (Binomial test, exact signification p:0.003) (Tables 4 and 5).

The width of the MN was 0.52 mm, SD 0.062 mm. No significant differences were observed depending on sex or side, although the width of the nerve in females was greater than in males. The diameter of the nerve did not vary in relation to the distance, either when measured by US (t Studdent −0.597, $p = 0.553$) or by anatomic study (t Studdent −0.934, $p = 0.354$).

When compared by the Bland–Altman plot, the agreement between the US and anatomy observations in all the measurements taken indicated that US always overestimates the measurements.

The tendon of the PL is inserted at the PA; however, we observed in 100% of the cases that this tendon expanded to the thenar muscles of the hand (Figure 6).

We found anatomical variations in 17% of the cases. Variations in the PL were detected by US in 98.61% of cases. Only in one variation of type 1 did we not detect it by US.

Variation type 0: US examination detected the absence of the PL, and anatomical study proved it in 11 upper limbs (15.27%)

Variation type 1: (two females) PL had no muscular fibers. It was only a tendon and was detected in 2.77% of cases.

https://doi.org/10.3390/app16020817

This was detected in one case by US and in two cases by the anatomical study. In the first, the US image was a slim hypoechogenic structure that originated at the ME along with the rest of the ME muscles' origin and was inserted at the PA (Figure 8A). Also, it was surrounded and isolated by the AF in the distal third of the forearm. The width of the tendon as measured by US was 0.25 mm superiorly and 0.21 mm inferiorly. Dissection allowed us to measure the entire tendon, which was 31 cm in length; the width was 0.20 mm superiorly and 0.15 inferiorly (Figure 8B).

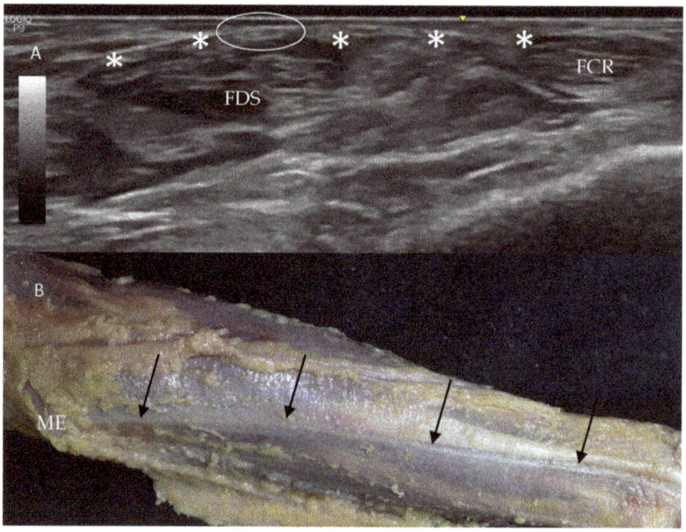

Figure 8. Variation type 1: (**A**): Ultrasound study showed a hypoechogenic structure (white circle) compatible with a tendon just under the antebrachial fascia (white *). (**B**): Anatomical dissection showed, after removing the skin and adipose tissue, a variation of type 1, with only a tendon of the palmaris longus muscle (black arrows). It runs from the medial epicondyle (ME) to the palmar aponeurosis without muscular fibers, covered by the antebrachial fascia.

The second case, only detected by dissection, revealed the presence of a tendon that was inserted at the PA; however, the tendon, with a length of 15.5 cm and width of 0.20 mm, originated from the medial side of the FCR (Figure 9) Moreover, there was a very slim tendon that came from the AF and joined with it before insertion (Figure 9).

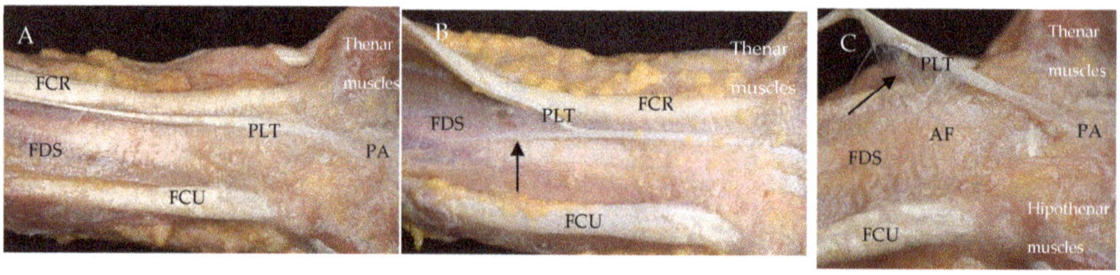

Figure 9. Variation types by anatomical study (**A**): The superficial layer of the antebrachial fascia is removed. The tendon of the PL (PLT) comes from the medial border of the flexor carpi radialis muscle (FCR) and inserts at the palmar aponeurosis (PA). (**B**): It is possible to see another tendon (black arrow) coming from the antebrachial fascia and joining with the PLT (black arrow). (**C**): The deep side of the PLT provides many connections to the antebrachial fascia (AF) (black arrow) before it inserts at the (PA). The flexor digitorum superficialis muscle (FDS) is covered in all the pictures by the antebrachial fascia. Flexor carpi ulnaris (FCU).

https://doi.org/10.3390/app16020817

Variation types 2 and 3 were found as hypo- or hyperechogenic structures at the medial and distal parts of the forearm, respectively, medial to the FCR, and covering the MN.

Variation type 2 (one female and one male): The muscular bell of the PL was in the middle of the forearm. The Yorigin and the insertion were a tendon. This was detected in 2.77% of cases. US images showed a broad hypoechogenic structure bigger than a normal tendon at the medial level of the forearm, similar to a normal PL, even with the intramuscular tendon that finishes with a hypoechogenic and wide structure. Dissection showed in one case that the PL originated from a tendon of 6 cm in length and 0.2 mm in width, followed by the muscular bell that had 16 cm of length and 1 cm of width. The AF surrounded the PL throughout its trajectory until it inserted at the PA (Figure 10). MN width was 0.5 cm. In some cases, the length of the tendon was 15.3 cm from the ME, followed by a muscular portion of 7.7 cm that finished at the PA with a tendon of 2.5 cm. The width of the tendon at the proximal was 0.5 mm, and at the distal part was 0.2 mm (Figure 11). The width of the MN was 0.55 mm by US and 0.52 by anatomical study.

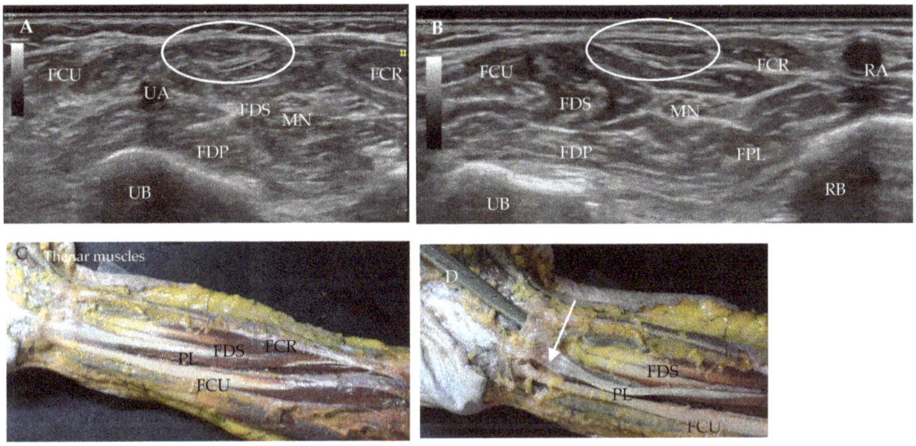

Figure 10. Variation type 2: (**A**): Ultrasound study at 9 cm from the medial epicondyle. PL has a typical muscular echogenicity and an ovoid shape (white circle). It is covered by the antebrachial fascia. It keeps its relationship with the flexor carpi radialis muscle (FCR) and the flexor carpi ulnaris muscle (FCU). Median nerve (MN), flexor digitorum superficialis (FDS), Flexor digitorum profundus (FDP), ulnar bone (UB). (**B**): Ultrasound image at 12 cm from the medial epicondyle is compatible with variation type 2. (**C**): Dissection of the forearm after removing the antebrachial fascia shows the central, long muscle of the PL with a tendinous portion. (**D**): Insertion of the distal tendon of PL (white arrow) under the antebrachial fascia (AF). Medial epicondyle (ME).

Variation type 3 (2 females): Also known as the inverted or reverse palmaris longus muscle, this variation was found in two cases (2.77%). The muscle was in the middle and distal parts of the forearm. In one case, the muscle had two origins from the ME as a wide and thin tendon, and the other, from the AF, along with the lacertus fibrosus. Muscular fibers originating from the PL were found at 8.6 cm from the ME. The length of the muscle was 15 cm, the width was 1.5 cm in the middle of the forearm, and at 2 cm from the FR. The lateral bell was inserted at the PA, and the medial bell was inserted at the medial side of the PA (Figure 12).

https://doi.org/10.3390/app16020817

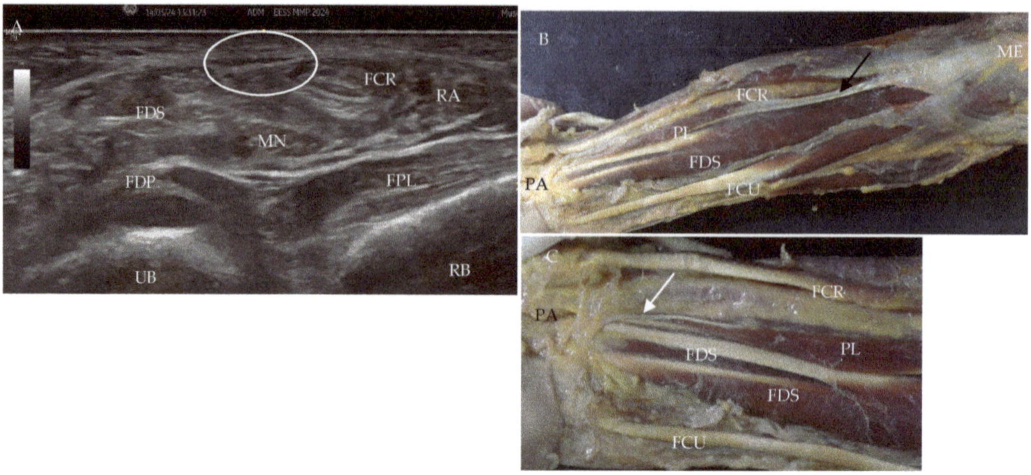

Figure 11. Variation type 2. (**A**): Ultrasound study of the middle of the forearm shows the palmaris longus muscle (PL) as a superficial hypoechogenic structure covered (white circle) by the antebrachial fascia. It keeps the relationship with the flexor carpi radialis muscle (FCR) and the flexor carpi ulnaris muscle (FCU). The median nerve (MN), flexor digitorum superficialis (FDS), Flexor digitorum profundus (FDP), flexor pollicis longuis (FPL), radius bone (RB), and ulnar bone (UB). (**B**): Dissection of the forearm after removing the antebrachial fascia shows the proximal muscle of the PL (black arrow) (**C**): Insertion of the distal and narrow tendon of PL (white arrow). Medial epicondyle (ME), palmar aponeurosis (PA).

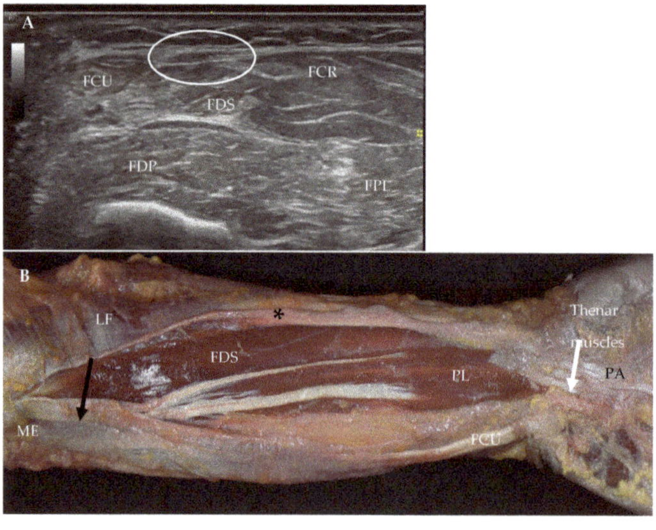

Figure 12. Variation type 3. (**A**): Ultrasound in the middle of the forearm detects the hypoechogenic image (white circle). Flexor carpi radialis (FCR), flexor carpi ulnaris (FCU), flexor digitorum superficialis muscle (FDS), flexor digitorum profundus (FDP), flexor pollicis longus (FPL). (**B**): An anatomical study of the forearm after the lacertus fibrosus (LF) and the antebrachial fascia (black *) were cut. The PL has two points of origin; one is lateral and originates at the medial epicondyle (ME), and the other is medial and originates at the antebrachial fascia near the ME (black arrow). The medial bell inserts at the medial side of the palmar aponeurosis (PA) (white arrow), and the lateral bell inserts directly at the palmar aponeurosis (PA).

The innervation of the bell came from the ulnar nerve. The diameter of the MN was 0.65 mm by US and 0.62 mm as measured by the anatomical study.

In the other case, the muscle bell of PL arrived at the PA. The length of the proximal tendon was 9 cm, and the length of the muscle was 16 cm; the width at the middle of the muscle was 1 cm. Additionally, the PL coexisted with an additional abductor digiti minimi muscle going under the ulnar vessels and the other over the ulnar nerve. The distance from the MN to the SP of the radius and the PL to this process was the same, 2 cm. The diameter of the MN was 0.61 mm as measured by US and 0.59 mm as measured by the anatomical study.

3.2. Other Variations

We found two cases with a double PL, one below the other. US imaging permitted the detection of these variations. These also originated at the ME with muscular fibers that followed distally and finished in a narrow tendon parallel to the PL tendon, which inserted at the AF or FR. Although the normal PL was considered in the statistical analysis, these two variations were an additional PL with independent origin and insertion points that were excluded, as they were not included in any of the described variation types. In fact, this additional PL was covered by the AF along its course.

3.3. Histology Study

Histological study showed the continuity of the tendon with the muscular fibers, with a very close relationship (Figure 13). The study of the tendon of the PL revealed a regular and typically dense connective tissue. The width of the tendon was 0.29 mm (Figure 13).

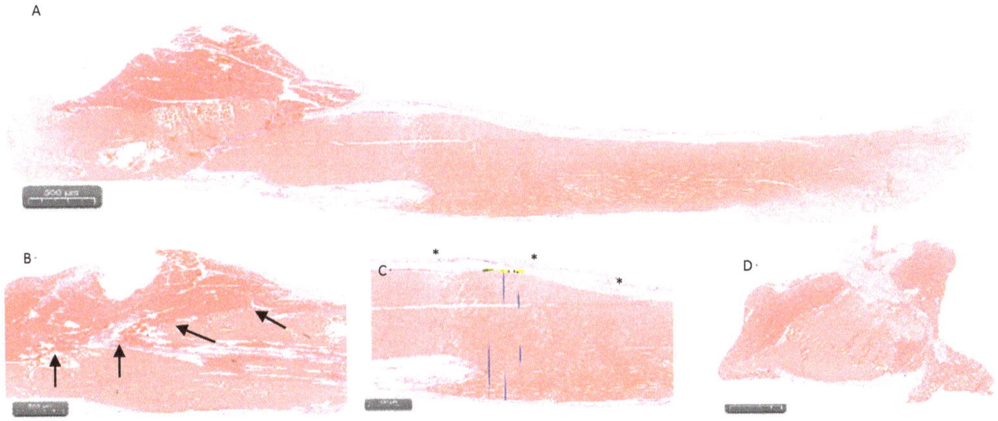

Figure 13. Histological view of the muscle and the tendon of the palmaris longus muscle (PL). (**A**): Longitudinal view of the end of the muscular portion (dark pink) and the tendon inside the muscle. (**B**): A view of the deep relationship between the tendon and the muscular fibers (black arrow) of the PL. The tendinous fibers are perfectly parallel. (**C**): Superior part of the tendon covered by the antebrachial fascia (black *). It has connections with the epitenon. Measures of the tendon were 0.29 mm. (**D**): Transversal view of the muscle (dark pink) and tendon at the end of the muscular portion.

4. Discussion

The PL has been extensively described by numerous authors with regard to its anatomy and variations [6,18,19] as well as its function [6,20,21]. The PL can be identified by a classical test (Schaeffer's test) by bringing the thumb and small fingertips together (forceful opposition of the thumb) and performing a mild active flexion of the wrist joint [12].

https://doi.org/10.3390/app16020817

However, the present study gives special importance to including an ultrasound study to detect the presence of the PL or its possible variations for their surgical and clinical applications and for the study of other carpal pathologies.

In agreement with previous research, our findings confirm that US is a reliable method for visualizing the PL and identifying its variants [22,23]. The US examination should begin at the ME, where a small triangular structure can typically be observed [4,22,24]. Nevertheless, it is important to move slightly distally to recognize the antebrachial fascia (AF), which appears as a hyperechogenic line and serves as an excellent reference for locating the PL. At this level, the AF clearly separates the PL from the FCR and FCU [4], allowing for their distinction in the proximal forearm. And also, to observe, as the probe is advanced distally, how the PL progressively changes its morphology from a triangular to an ovoid shape, with a typical central hyperechogenic line corresponding to the origin of the tendon of the PL. The anatomical dissection corroborates these ultrasonographic features, particularly the fusiform configuration previously reported by other authors [6,10]. Importantly, our anatomical observations emphasize that the PL originates from the ME by means of a tendinous structure, rather than by direct muscular fibers [16]. The muscle fibers do not extend to the ME itself.

The measured length of the muscular portion of the PL (12.01 $\pm$ 1.708 mm) was slightly shorter than that reported in previous studies [20], although it remained within the expected anatomical range. This can be explained by the differences between populations or even different references about the limits of the muscle.

With respect to the length of the tendon of PL (13.8 $\pm$ 2.4 cm), it was very similar to other studies [14,25]. The width of the tendon (0.39 mm), sometimes, was slightly bigger at the proximal point than the distal point as measured by US, and was also similar to some other studies [14,23] but quite different from other studies (0.88 mm) performed with living people, which may have measured at the level of the carpal bones [22]. The histological measures, although they were less than those observed with the US, and the anatomical results were very similar. Interestingly, no significant differences were observed between sexes or sides in any of the measurements obtained in this study. These measures can help with surgical applications seeking to use the PL tendon.

The histological structure of the PL tendon was consistent between the middle and distal thirds, as previously reported by other authors [20]. The tendon maintains a uniform morphology throughout its course, although the presence of abundant and large nerve fibers has been documented [26]. We have not studied them, but they can be important for movement. Another aspect to consider is the deep relation of the muscle fibers and the tendon. This could be important to preserve the function of the muscle in case of surgery, as several studies indicate that it may cause a decrease in the function of the palmar flexion [27]. Its uniformity and its length prove that it has several advantages. The most important, and in accordance with the main objective of this study, is that it is valuable for reconstruction and tendon grafting procedures [12,13]. However, we think it is mandatory to conduct a US study to prove its existence.

Of particular relevance is the relationship between the PL tendon and the AF, which has both functional and surgical implications. As demonstrated by our US and anatomical observations, the PL tendon consistently descends covered by the AF [3,20]. This relationship varies slightly between the middle and distal thirds of the forearm, as also described by other authors [4]. Such differences should be considered during surgical procedures. We recommend accessing the distal third of the forearm and incising only the sheath of the PL up to the level of the tendon. This approach preserves the AF that envelops the remaining anterior forearm muscles, as well as the MN and the ulnar neurovascular bundle, thereby minimizing the risk of iatrogenic injury. It is essential to preserve muscular function and

https://doi.org/10.3390/app16020817

ensure MN protection. Conversely, harvesting the PL tendon at the distal and anterior wrist crease [28,29] is not advisable, as it carries a higher risk of MN injury.

The relevance of the AF is further illustrated by other anatomical variants in which a double PL is observed; in these cases, the accessory PL inserts into the AF, potentially contributing to its tension and functional stability.

Our study also quantified the distance between the MN and the PL tendon relative to the SP of the radius bone, which provided a clear understanding of their anatomical relationship at the wrist. This information is especially useful during surgery, as the anatomical superficial PL tendon and medial position will help to distinguish the MN. However, we assert that a previous US study can be very useful to distinguish them. In many specimens, the MN was partially or completely overlain by the PL tendon, a finding that could help to explain the chronic compression of the MN caused by repetitive wrist movements, a mechanism potentially contributing to carpal tunnel syndrome [30]. Nevertheless, we do not believe that the PL tendon itself is a direct compressive factor on the MN. In our data, the average MN width was 0.55 mm, closely matching previous reports (0.59 $\pm$ 0.93 mm) [4], irrespective of tendon–nerve proximity. Additionally, no significant differences were observed between sexes or sides. We suggest that the AF, by separating the PL tendon from the MN, acts as a protective barrier, reducing potential compression. However, in type 3 variations, where muscular fibers extend distally and lie directly over the MN, contraction of these fibers may exert pressure on the nerve. In fact, in our series, the MN width was slightly greater in these cases, supporting this hypothesis.

As described in other studies, we also observed that the insertion of the PL tendon is consistently located anterior to the FR, attaching primarily to the PA and occasionally to other structures such as the AF, FR, thenar, and hypothenar regions, or carpal bones including the trapezium, scaphoid, and pisiform [7], as well as the tendons of the FCU, FCR, FDS, and FDP [16]. Despite the abundance of connective tissue in both the PL tendon and the PA, they do not share a common origin [31], which explains the wide variability of connections between them [32]. The PL appears to insert into the subcutaneous plane of the palm, continuing into the superficial layer of the PA formed by its longitudinal fibers. This anatomical configuration supports the interpretation of the PL as a tensor of the superficial fascial system of the palmar subcutaneous tissue [20].

Performing an US evaluation prior to any surgical procedure is therefore essential for detecting PL variations, given that it is among the most variable muscles in the human body [33]. US findings can guide both the feasibility assessment and surgical strategy for reconstructive interventions.

Previous authors have classified PL variations based either on tendon insertion morphology or on the full spectrum of anatomical variants [6,18]. In this study, however, we propose a more functional and clinically oriented classification, applicable to both US and surgical practice. This simplified system considers the position of the muscular belly and the insertion pattern at the PA, facilitating intraoperative identification and planning [22].

The type 0 variation, or agenesis of the PL, is the most frequent variant [22] and our findings are consistent with those of previous studies. This absence has been attributed to Mendelian inheritance [21] and varies not only among ethnic groups but also among individuals within the same population [21]. No statistically significant differences were found regarding sex or laterality [22], although other authors have reported a higher incidence in females and on the left side [1,8]. However, others report a higher incidence in males [12,34] as well as a greater frequency of bilateral rather than unilateral agenesis [22]. In general, an absence rate of approximately 10% has been reported [12,18,35], while we observed a slightly higher incidence of 15.27%, comparable to that found by other authors [8,22]; some studies even report rates as high as 25% [7]. Recognizing this type of

https://doi.org/10.3390/app16020817

variation is crucial to prevent misidentification of the MN during surgery, given its close anatomical proximity [35].

Type 1 variation in the PL was identified in 2.77% of the specimens. No previous studies were found in the literature for direct comparison. This variant is easily recognizable by US, as the tendon exhibits the same echogenic characteristics as the typical PL tendon, although it is slightly thinner—approximately 0.2 mm narrower. Despite its reduced caliber, we consider that this variation may still be suitable for use in reconstructive surgical procedures.

The type 2 variation was also observed in 2.77% of cases, although the two specimens presented distinct morphologies. In this variant, a proximal tendon is followed by a central muscular belly and a distal tendon [6,10]. Another possible configuration is the digastric PL, characterized by two muscular bellies [6]; these may insert independently into the PA [6], or both may converge into a single tendon that subsequently inserts into the PA [10]. In some reports, a trigastric PL has also been described, consisting of three muscular bellies with separate tendons that all insert into the PA [6], though such forms were not identified in our series. From a surgical standpoint, type 2 variations may be clinically valuable, particularly when the distal tendon is sufficiently long to be used in reconstructive procedures. In these cases, the PL tendon could be suitable for surgical grafts.

Type 3 variation is perhaps the most clinically significant variation in the PL and should be carefully assessed using US. This variant, also known as the reversed or inverted palmaris longus [6,36–39], presents a muscular belly in the distal portion of the forearm, whose morphology can vary considerably. Some authors include within this category the bifid PL [6,10] or cases with multiple muscular heads [5,40]. However, we emphasize that this type should not be confused with instances in which the muscular fibers reach the PA with a tendon, although the tendon is short. In such cases, it would be classified as a type 2 variation. We determined that this variation is not suitable for surgical treatments. Alternatively, in these variations, perhaps surgery could be indicated to avoid the compression of MN, if it is present.

In our classification, we also included cases of reversed PL coexisting with the abductor digiti minimi muscle, as previously described by other authors [6], although in our specimens, the PL inserted into the hypothenar musculature. Other authors describe the abductor digiti minimi muscle in relationship with the PL tendon. We emphasize that it is important to consider the possibility of compression of the ulnar nerve or vessels, as this could complicate surgical procedures requiring resection of the palmaris longus tendon [41,42]. Recognizing this variation is important because it is unsuitable for reconstructive surgery, given the distal position of its muscular belly and its limited tendinous length. Moreover, it carries diagnostic relevance, as it may be associated with pain in the volar aspect of the distal forearm due to compression of the MN or, less frequently, of the ulnar nerve and vessels [41]. Repetitive wrist movements may lead to hypertrophy of the distal muscle belly; owing to the rigidity of the AF, this can precipitate an effort-related compartment syndrome. In such cases, US proves highly valuable for accurate identification and differential diagnosis of this condition.

It is also worth emphasizing that the PL, owing to its high proportion of connective tissue, plays a significant role in the wrist joint proprioception, fine motor skills, and nociceptive functions of the forearm and hand. However, the last investigations confirm that the presence or absence of the PLM does not affect fine motor skills, grip strength, or wrist proprioception, suggesting it can be safely used for surgical grafting without functional loss in the hand [27]

Consequently, its morphology should be regarded as dynamic rather than regressive, reflecting an anatomical evolution more oriented toward sensory integration than pure motor performance (a transformation in structure driven by a change in function). Adopting a more holistic perspective on the anatomophysiological role of the PL could therefore expand our understanding of forearm function and contribute to improved clinical approaches to specific pathologies affecting this region.

5. Conclusions

Comprehensive knowledge of the anatomy and possible variations in the palmaris longus muscle is essential, given its importance in reconstructive surgical procedures, as the tendon of PL can be the main option for autograft tendon transfer in different areas of the body. However, it is necessary to prevent surgical complications or misidentification through a thorough understanding of its anatomical variability. For this reason, we believe it is mandatory to conduct a preoperative ultrasound assessment. This is the best technique, and is therefore indispensable for identifying the presence of the palmaris longus muscle and its variations, and ensuring safe and accurate surgical planning. Some of these variations could also be useful for surgical procedures.

Author Contributions: Conceptualization, M.M.-P., A.M.D. and A.P.-B.; Methodology, M.M.-P. and A.M.D.; Software, A.M.D., I.M. and C.M. (Carlo Martinoli); Validation, M.M.-P., A.M.D. and C.M. (Carlo Martinoli); Formal analysis, J.C.O.-S.; Investigation, M.M.-P., A.M.D., S.O.-M., S.L.-P., N.R.-R., I.M. and A.P.-B.; Resources, A.M.D., S.O.-M., S.L.-P., N.R.-R. and I.M.; Data curation, A.M.D., S.O.-M., J.C.O.-S., C.M. (Cristina Miralles), N.R.-R., I.M. and A.P.-B.; Writing—original draft, M.M.-P., S.O.-M. and A.P.-B.; Writing—review & editing, M.M.-P. and A.P.-B.; Visualization, M.M.-P., S.O.-M., S.L.-P. and C.M. (Cristina Miralles); Supervision, M.M.-P., S.O.-M., C.M. (Carlo Martinoli) and A.P.-B.; Project administration, M.M.-P. and S.O.-M.; Funding acquisition, M.M.-P. All authors have read and agreed to the published version of the manuscript.

Funding: This research received no external funding.

Institutional Review Board Statement: The study was conducted in accordance with the Declaration of Helsinki, and approved by the Institutional Review Board IRB00003099, signed 28th of Ferber of 2022).

Informed Consent Statement: Informed consent was obtains from all subjects involved in the study in life.

Data Availability Statement: The original contributions presented in this study are included in the article. Further inquiries can be directed to the corresponding author.

Acknowledgments: To all our donors.

Conflicts of Interest: The authors declare no conflict of interest.

Abbreviations

AF	Antebrachial fascia
FCR	Flexor carpi radialis
FCU	Flexor carpi ulnaris
FR	Flexor retinaculum
ME	Medial epicondyle of humerus bone
MN	Median nerve
PA	Palmar aponeurosis
PL	Palmaris longus
SP	Styloid process
US	Ultrasound

https://doi.org/10.3390/app16020817

References

1. Javaid, Q. Prevalence of agenesis of palmaris longus in male and female genders on clinical examination—A review. *J. Pak. Med. Assoc.* **2022**, *72*, 733–737.
2. Khale, W.; Leonbardt, H.; Platzer, W. *Atlas de Anatomía. Aparato Locomotor. Tomo 1*; Ediciones Omega, S.A.: Barcelona, Spain, 1995; 434p.
3. Schünke, M.; Schulte, E.; Schumacher, U.; Voll, M.; Wesker, K. *Prometheus-Texto y Atlas de Anatomía-Tomo 1-Anatomía General y Aparato Locomotor*; Panamericana, M., Ed.; Editorial Médica Panamericana S.A.: Buenos Aires, Argentina, 2016; 628p.
4. Ortiz-Miguel, S.; Miguel-Pérez, M.; Navarro, J.; Möller, I.; Pérez-Bellmunt, A.; Agullo, J.L.; Ortiz-Sagristà, J.; Blasi, J.; Martinoli, C. Compartments of the antebrachial fascia of the forearm: Clinically relevant ultrasound, anatomical and histological findings. *Surg. Radiol. Anat.* **2021**, *43*, 1569–1579. [CrossRef]
5. Sunil, V.; Rajanna, S.; Gitanjali; Kadaba, J. Variation in the insertion of the palmaris longus tendon. *Singap. Med. J.* **2015**, *56*, e7–e9. [CrossRef]
6. Georgiev, G.P.; Iliev, A.A.; Dimitrova, I.N.; Kotov, G.N.; Malinova, L.G.; Landzhov, B.V. Palmaris Longus Muscle Variations: Clinical Significance and Proposal of New Classifications. *Folia Medica* **2017**, *59*, 289–297. [CrossRef]
7. Bernardes, A.; Melo, C.; Pinheiro, S. Une combinaison de variations des muscles Palmaris longus et Flexor digitorum superficialis: à propos d'un cas et revue de la littérature. *Morphologie* **2016**, *100*, 245–249. [CrossRef]
8. Javaid, Q.; Usmani, A.; Surti, A. Anatomical aberration of palmaris longus agenesis muscle through clinical examination in relation to gender and hand dominance. *J. Pak. Med. Assoc.* **2023**, *73*, 74–77.
9. Garbarino, M.C.; Pisani, A.; Biggiogera, M.; Mazzarello, P. Camillo Golgi's contributions to the anatomic basis of sensitivity in tendons. *J. Neural Transm.* **2024**, *132*, 287–293. [CrossRef]
10. Olewnik, Ł.; Waśniewska, A.; Polguj, M.; Podgórski, M.; Łabętowicz, P.; Ruzik, K.; Topol, M. Morphological variability of the palmaris longus muscle in human fetuses. *Surg. Radiol. Anat.* **2018**, *40*, 1283–1291. [CrossRef]
11. Ozturk, E.K.; Ozturk, S.; Taflan, S.S. Palmaris longus tendon and its relationship with the median nerve: A population-based morphometric magnetic resonance imaging study. *Quant. Imaging Med. Surg.* **2025**, *15*, 9125–9132. [CrossRef]
12. Sambhav, K.; Krishna, H.; Singh, P.; Bhardwaj, N.; Ghatak, S. Anatomical Variations of Palmaris Longus Muscle and Its Clinical Implications: A Cadaveric Study. *Maedica-A J. Clin. Med.* **2025**, *20*, 560–565. [CrossRef]
13. Gelaw, L.T.; Gebremariyam, Z.T.; Woldemicael, N.A.; Habtemariam, M.S. Pioneering Radial Forearm Free Flap Reconstruction for Lip Cancer in a Young Patient in Ethiopia. *Plast. Reconstr. Surg.–Glob. Open* **2025**, *13*, e7349. [CrossRef]
14. Yammine, K.; Erić, M. Morphometric analysis and surgical adequacy of palmaris longus as a tendon graft. A systematic review of cadaveric studies. In *Surgical and Radiologic Anatomy*; Springer: Berlin/Heidelberg, Germany, 2020; Volume 42, pp. 259–267.
15. Fritz, T.; Ducommun, P.; Pohlemann, T.; Calcagni, M.; Tschernig, T.; Menger, M.D.; Metzger, W.; Frueh, F.S. Flexor tendon grafts for pulley reconstruction–Morphological aspects. *Ann. Anat.* **2020**, *231*, 151550. [CrossRef]
16. Yuan, J.; Dhanaraj, I.D.; Ng, H.J.H.; Rajaratnam, V. Accessory Palmaris Longus: A Concise Anatomic Review with a Relevant Case Study. *Ann. Plast. Surg.* **2019**, *83*, 363–366. [CrossRef]
17. Moore, C.W.; Fanous, J.; Rice, C.L. Fiber type composition of contiguous palmaris longus and abductor pollicis brevis muscles: Morphological evidence of a functional synergy. *J. Anat.* **2021**, *238*, 53–62. [CrossRef]
18. Olewnik, Ł.; Wysiadecki, G.; Polguj, M.; Podgórski, M.; Jezierski, H.; Topol, M. Anatomical variations of the palmaris longus muscle including its relation to the median nerve-A proposal for a new classification. *BMC Musculoskelet. Disord.* **2017**, *18*, 539. [CrossRef]
19. Martinoli, C.; Perez, M.M.; Padua, L.; Valle, M.; Capaccio, E.; Altafini, L.; Michaud, J.; Tagliafico, A. Muscle variants of the upper and lower limb (with anatomical correlation). *Semin. Musculoskelet. Radiol.* **2010**, *14*, 106–121. [CrossRef]
20. Stecco, C.; Lancerotto, L.; Porzionato, A.; Macchi, V.; Tiengo, C.; Parenti, A.; Sanudo, J.R.; De Caro, R. The palmaris longus muscle and its relations with the antebrachial fascia and the palmar aponeurosis. *Clin. Anat.* **2009**, *22*, 221–229. [CrossRef]
21. Pai, M.M.; Prabhu, L.V.; Nayak, S.R.; Madhyastha, S.; Vadgaonkar, R.; Krishnamurthy, A.; Kumar, A. The palmaris longus muscle: Its anatomic variations and functional morphology. *Rom. J. Morphol. Embryol.* **2008**, *49*, 215–217.
22. Fidan, N.; Turk, A.C.; Yetis, E.U.M.; Yucesoy, C. Determination of congenital absence of palmaris longus tendon with clinical examination and ultrasonography. *Surg. Radiol. Anat.* **2021**, *43*, 741–748. [CrossRef]
23. Johnson, C.C.; Vutescu, E.S.; Miller, T.T.; Nwawka, O.K.; Lee, S.K.; Wolfe, S.W. Ultrasound determination ofpresence, length and diameter of the palmaris longus tendon. *J. Hand Surg. Eur. Vol.* **2018**, *43*, 948–953.
24. Şahin, Z.A.; Üzel, M.; Yıldırım, F.G.K.; Tanyeli, M.E.; Ertaş, A.; Palamar, D.; Mısırlıoğlu, T.Ö.; Akgün, K.; Dirican, A. Sensitivity and specificity assessment of four clinical tests for palmaris longus muscle by ultrasound. *Turk. J. Phys. Med. Rehabil.* **2024**, *70*, 525–531. [CrossRef]
25. Ito, M.M.; Aoki, M.; Kida, M.Y.; Ishii, S.; Kumaki, K.; Tanaka, S. Length and width of the tendinous portion of the palmaris longus: A cadaver study of adult Japanese. *J. Hand Surg.* **2001**, *26*, 706–710. [CrossRef]

https://doi.org/10.3390/app16020817

26. Jozsa, L.; Balint, J.; Kannus, P.; Lehto, M. Mechanoreceptors in human myotendinous junction. *Muscle Nerve* **1993**, *16*, 453–457. [CrossRef]
27. Sahin, N.E.; Uysal, M.; Sahin, M.; Toy, S.; Ozmen, T. The effect of the palmaris longus muscle on fine motor skills, position sense and grip strength. *J. Back Musculoskelet. Rehabil.* **2025**, *24*, 10538127251380061. [CrossRef]
28. Jakubietz, M.G.; Jakubietz, D.F.; Gruenert, J.G.; Zahn, R.; Meffert, R.H.; Jakubietz, R.G. Adequacy of palmaris longus and plantaris tendons for tendon grafting. *J. Hand Surg.* **2011**, *36*, 695–698. [CrossRef]
29. Tomori, Y.; Nanno, M.; Kentaro, S.; Majima, T. Novel surgical procedure for half palmaris longus transfer during opponensplasty of the thumb for patients with carpal tunnel syndrome: A technical note. *J. Nippon. Med. Sch.* **2021**, *88*, 149–153. [CrossRef]
30. Boltuch, A.D.; Marcotte, M.A.; Treat, C.M.; Marcotte, A.L. The Palmaris Longus and Its Association with Carpal Tunnel Syndrome. *J. Wrist Surg.* **2020**, *9*, 493–497. [CrossRef]
31. Caughell, K.A.; McFarlane, R.M.; McGrouther, D.A.; Martin, A.H. Developmental anatomy of the palmar aponeurosis and its relationship to the palmaris longus tendon. *J. Hand Surg.* **1988**, *13*, 485–493. [CrossRef]
32. Fahrer, M. The proximal end of the palmar aponeurosis. *Hand* **1980**, *12*, 33–38. [CrossRef]
33. Gervasio, A.; Stelitano, C.; Bollani, P.; Giardini, A.; Vanzetti, E.; Ferrari, M. Carpal tunnel sonography. *J. Ultrasound* **2020**, *23*, 337–347. [CrossRef]
34. Al Risi, A.M.; Al Busaidi, S.; Al Aufi, H.; Al Hashmi, L.; Sirasanagandla, S.R.; Das, S. Anatomical Study of the Palmaris Longus Muscle and Its Clinical Importance. *Diagnostics* **2025**, *15*, 304. [CrossRef] [PubMed]
35. Woon, C.Y.L.; Gourineni, R.; Watkins, S.; Richardson, R.; Gourineni, P. That Tendon is the Median Nerve: Perils of a Palmaris Longus Deficient Wrist. *J. Hand Surg. Asian-Pac. Vol.* **2017**, *22*, 160–166. [CrossRef]
36. Nayak, S.B.; Packiriswamy, V.; Vasudeva, S.K. An unreported variant of palmaris longus muscle. *Anat. Cell Biol.* **2024**, *57*, 605–607. [CrossRef]
37. Behiery, A.; Bakir, M.; Dawalibi, A.; Elhossiny, A.H. Reversed Palmaris Longus: A Rare Anatomical Phenomenon Discovered. *Cureus* **2023**, *15*, e38424. [CrossRef] [PubMed]
38. Yildiz, M.; Sener, M.; Aynaci, O. Three-headed reversed palmaris longus muscle: A case report and review of the literature. *Surg. Radiol. Anat.* **2000**, *22*, 217–219. [CrossRef] [PubMed]
39. Longhurst, G.; Stone, D.; Mahony, N. Bilateral reversed palmaris longus muscle: A case report and systematic literature review. *Surg. Radiol. Anat.* **2020**, *42*, 289–295. [CrossRef]
40. Raouf, H.A.; Kader, G.A.; Jaradat, A.; Dharap, A.; Fadel, R.; Salem, A.H. Frequency of palmaris longus absence and its association with other anatomical variations in the egyptian population. *Clin. Anat.* **2013**, *26*, 572–577. [CrossRef]
41. Murabit, A.; Gnarra, M.; Mohamed, A. Reversed palmaris longus muscle: Anatomical variant-case report and literature review. *Can. J. Plast. Surg.* **2013**, *21*, 55–56. [CrossRef]
42. Hamacher, C.D.; Browning, B.J.; Barton, J.C.; Chase, N.F.; Larson, R.E.; Green, D.J. Presence of accessory abductor digiti minimi muscle in two cadavers. *Folia Morphol.* **2023**, *82*, 216–220. [CrossRef]

https://doi.org/10.3390/app16020817

applied sciences

MDPI

Article

Plane Wave Imaging with Large-Scale 2D Sparse Arrays: A Method for Near-Field Enhancement via Aperture Diversity

Óscar Martínez-Graullera *, Jorge Camacho, Jorge Huecas, Guillermo Cosarinsky, Luis Elvira and Montserrat Parrilla

Instituto de Tecnologías Físicas y de la Información (ITEFI), Consejo Superior de Investigaciones Científicas (CSIC), 28006 Madrid, Spain; j.camacho@csic.es (J.C.); jorge.huecas@csic.es (J.H.); g.cosarinsky@csic.es (G.C.); luis.elvira@csic.es (L.E.); m.parrilla@csic.es (M.P.)

* Correspondence: oscar.martinez@csic.es

Abstract: In the context of a medical imaging application for preclinical research, specifically cerebrovascular imaging in small animals, this work addresses the challenges associated with using a large-scale 2D ultrasonic array comprising 32 × 32 elements ($96\lambda \times 96\lambda$). The application imposes stringent requirements: operation in the extreme near field, high spatial resolution, high frequency, high frame rate, and imaging within a highly attenuating medium. These demands, combined with current technological limitations, such as element size and constraints on the number of channels that can be driven in parallel, present significant challenges for system design and implementation. To assess system performance, plane wave imaging is employed as a reference modality due to its ability to meet high acquisition speed requirements. Our analysis reveals limitations in spatial coverage and image quality when operating the full aperture under plane wave transmission constraints. To address these limitations, we propose a sparse aperture strategy. When combined with advanced signal processing techniques, this approach enhances both contrast and resolution while preserving acquisition speed, making it a promising solution for high-performance ultrasonic imaging under the demanding conditions of preclinical research.

Keywords: ultrasonic imaging; bidimensional aperture; ultrafast imaging; beamforming

1. Introduction

Ultrasound imaging has become an essential tool in preclinical cerebrovascular research due to its ability to provide non-invasive, real-time, and high-resolution visualization of cerebral anatomy and hemodynamics [1,2]. It enables assessment of cerebral blood flow, tissue perfusion, and blood–brain barrier integrity in animal models, supporting studies of ischemic stroke, cerebral hemorrhage, and small vessel disease [3]. Contrast agents enhance sensitivity to microvascular changes and vascular permeability. It is also used to monitor therapeutic responses, including neuroprotective agents and reperfusion strategies. Its affordability and portability enable longitudinal studies without repeated invasive procedures.

When combined with other modalities such as MRI or PET [4], ultrasound supports multimodal analysis, providing a flexible and efficient platform for studying cerebrovascular pathophysiology and developing novel therapies [1,2,5,6].

However, despite these advantages, implementing high-frequency ultrasound imaging in animal research presents several technical challenges [7–10]. While higher frequencies are essential for resolving small anatomical structures, such as the microvasculature in rodent brains, they are subject to increased acoustic attenuation in biological tissues, which limits penetration depth and degrades signal quality in deeper regions. Anatomical access is further constrained by the skull [11], which attenuates ultrasound waves and often necessitates craniotomies or acoustic windows to maintain image quality [12]. These limitations are especially critical in volumetric imaging with 2D arrays, which enables three-dimensional reconstruction of vascular flow and morphology but is highly sensitive to motion artifacts—such as respiration and cardiac pulsation—and remains restricted to superficial cortical regions.

These limitations highlight the need for optimized imaging protocols, specialized transducer designs, and advanced compensation strategies to improve image quality and enhance the utility of ultrasound in preclinical cerebrovascular studies [8,12–15].

In this context, plane wave imaging has emerged as a key technique for enabling ultra-fast acquisitions and efficient use of large apertures. By transmitting unfocused waves across the entire array, this method allows for rapid capture of cerebral blood flow dynamics with high temporal resolution [16].

Although plane wave imaging can generate an image from a single transmission, it tends to produce significant sidelobes and reduced image quality compared to focused beam techniques. These artifacts can be mitigated by compounding multiple images acquired at different transmission angles, thereby enhancing overall image fidelity. In linear arrays, typical angle sets range from 3 to 30, while 2D arrays commonly use configurations spanning from 5×5 to 11×11 angles. Determining the optimal number of transmission angles requires experimental validation or simulation to balance image quality and acquisition complexity [9,12,17].

However, volumetric imaging still presents significant technological challenges in terms of aperture design. Achieving apertures large enough to ensure high spatial resolution is difficult, especially at high frequencies. Maintaining half-wavelength element spacing to avoid grating lobes conflicts with the need to transmit sufficient acoustic energy into the medium and to ensure adequate element sensitivity. This constraint forces the use of very small elements, which typically exhibit low radiation efficiency and high electrical impedance, further complicating transducer design and signal transmission. As a result, the number of parallel channels required often exceeds the capabilities of current electronic systems and the system tends to operate with low efficiency.

To address these challenges, there is a growing trend toward the development of transducers with larger elements [13,18]. Most research efforts rely on large-scale 2D ultrasonic arrays with 32×32 elements (15 MHz, BW = 70%), featuring inter-element spacing of approximately 3λ, which represents the largest commercially available aperture (Vermon S.A.). However, due to the element size, these systems have limited steering capabilities and are complex to operate. Moreover, controlling such a large number of elements in parallel is technically demanding. Nowadays, several systems that can manage up to 256 channels simultaneously have emerged, often relying on complex multi-system configurations. Unfortunately, this introduces incompatibilities with the plane wave imaging modality.

This work explores the implementation of a volumetric imaging system based on plane wave transmission using large apertures, while operating with a reduced number of active elements for preclinical imaging in small animal models. Specifically, using a 32×32 array with 3λ element spacing, we investigate the contribution of individual elements to image formation in the extreme near field, as well as their limitations in beam steering. Based on these findings, we propose a solution that employs 256 active elements to generate plane wave images by leveraging the spatial diversity of the transmit aperture, rather than relying on beam deflection, to enhance image quality. This work focuses on the emission part of the process, particularly on the suppression of artifacts caused by the flat front on the received signals. This approach also enables system-level optimizations that improve data encoding and reduce the data throughput between the acquisition system and the beamforming and processing unit.

2. Materials and Methods

The intended application focuses on stroke research and aims to develop analytical tools for tissue characterization, anatomical structure identification, and cerebrovascular system analysis in rat models (see Figure 1). The project is currently in its early stages, prior to experimental work, during which we are defining the instrumentation, evaluating its operability, and identifying the necessary adaptations for the specific application.

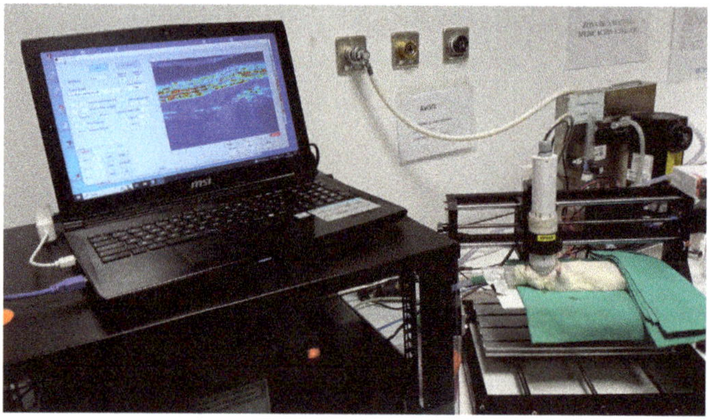

Figure 1. Prototype of an experimental high-frequency single-transducer ultrasound system, developed within the framework of project PID2022-138013OB-I00 (MCIN/AEI/10.13039/501100011033/FEDER, EU), which served as a precursor to the imaging approach proposed in this work.

The rat brain has an approximate volume of 1765 mm^3, with a length ranging from 2 to 2.5 cm and a depth between 1 and 1.5 cm. From an ultrasound perspective, it presents several challenges: high-frequency imaging is required to resolve small anatomical structures, the tissue is highly attenuating, and the skull acts as a barrier that is difficult to penetrate [19]. In this context, it is essential to position the transducer as close as possible to the region of interest to minimize attenuation effects. However, this may place the transducer in a region where its performance is suboptimal. Additionally, working with small animals introduces practical constraints. Although several studies address this type of application and describe positioning and coupling mechanisms, few provide details on array operability [20,21]. In anatomical imaging mode, image quality is limited, and large volumes of data combined with complex post-processing are often required to accurately reconstruct cerebrovascular structures.

In our case, we will be working with the SITAU II ultrasound imaging system (DASEL SISTEMAS S.L., Arganda del Rey, Spain), which supports 256 channels and operates with a two-dimensional aperture at 15 MHz. The array consists of 32 $\times$ 32 elements, with a lateral size of 96 wavelengths and a bandwidth of 70% (manufactured by Vermon S.A., Tours, France). The objective of this work is to analyze the performance of the aperture to determine its applicable range and to adapt it to a configuration of 256 active elements by designing the multiplexing system and evaluating its operability.

This study is conducted in a theoretical framework and is based on simulation models. For this purpose, custom computational tools have been developed using the simulation framework proposed by Piwakowski [22].

3. Large-Scale 2D Array

In a two-dimensional array configuration, using elements larger than $\frac{\lambda}{2} \times \frac{\lambda}{2}$ introduces significant design challenges, even in the relatively straightforward case of a full matrix array. Beyond balancing system performance with hardware constraints, such as reducing the total number of elements, the design process must also address complex trade-offs between the aperture's inherent physical characteristics and the specific operational requirements of the application. These requirements typically include penetration depth, beam-steering capability, spatial resolution, dynamic range, and individual element sensitivity.

In this context, a matrix array with elements sized at $3\lambda \times 3\lambda$ (see Figure 2A) clearly prioritizes radiation efficiency and sensitivity over beam-steering capability. Compared to half-wavelength elements, this configuration achieves a 36× increase in both radiation efficiency and sensitivity. However, this improvement introduces diffraction-induced modulation in the array response, resulting in a sensitivity drop of approximately −6 dB at an angle of 10°, thereby limiting the array's steering range. Figure 2B illustrates the focal behavior at various steering angles and highlights the associated reduction in dynamic range.

This design choice becomes evident when considering the attenuation characteristics of the inspection scenario. At 15 MHz, the acoustic attenuation within the biological tissue is approximately 7.5 dB/cm. Additionally, the use of a coupling medium introduces further attenuation of around 1.8 dB/cm. Consequently, maximizing the transmitted energy into the biological tissue is critical, and minimizing the propagation path through the coupling layer is advantageous. Figure 2C shows the impact of attenuation on the acoustic field along the propagation axis for the biological tissue and for the coupling layer.

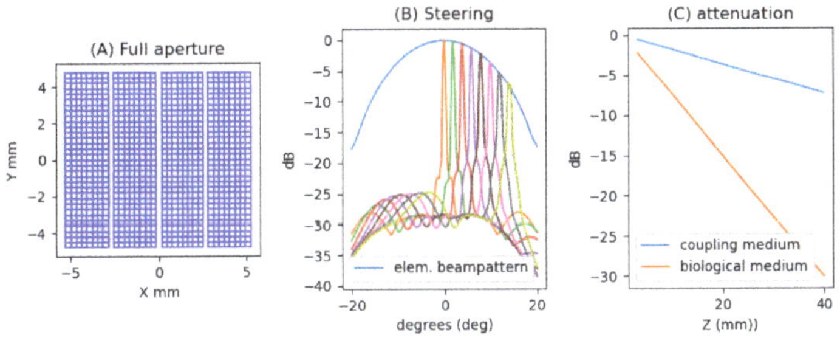

Figure 2. (**A**) Element configuration of the 32 × 32 array. (**B**) Lateral profile showing the steering capabilities of the full aperture, from 0° to 2°:14°, and evolution of the beam from −20° to 20°. (**C**) Evolution of the attenuation with the propagation in the coupling layer and in the biological tissue.

Based on the attenuation curves and, as an example, assuming that a coupling layer of at least 10 mm is required, the region of interest should not exceed 40 mm. Under these conditions, we would accept an attenuation of 22.5 dB between the beginning and the end of the sample. Considering a pulse–echo process, in 30 mm this results in a total loss of approximately 50 dB. This example illustrates the problem of attenuation and highlights the need for proper insonification of the medium in order to recover the maximum amount of reflected energy. From this point onward, the attenuation effect of the coupling medium will be included in all simulations.

3.1. Plane Wave Generation

Considering the limitations introduced by increasing the element size for conventional beam steering, we must study how this affects the process of deflecting a plane wavefront. In Figure 3, we present the beampatterns for several cases of planar aperture steering (at 0°, 6°, and 12°, see Figure 3A,B,D), and we show the cross-section at ($Z = 30$ mm, $Y = 0$ mm) for steering angles ranging from 0° to 14°.

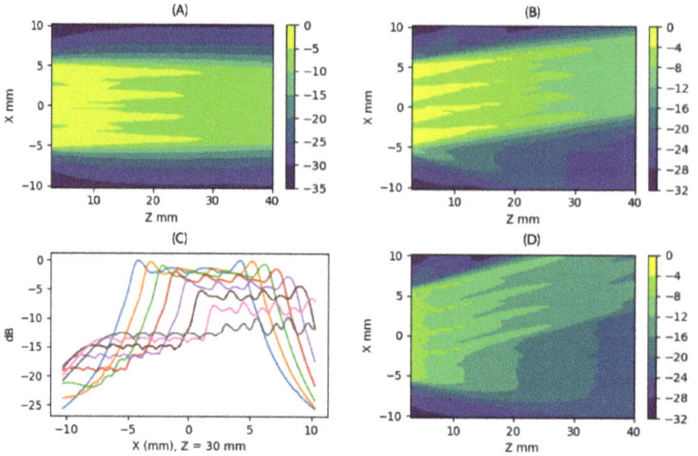

Figure 3. Beampattern for planar emission in near-field cases of (**A**) 0°, (**B**) 6°, and (**D**) 12°. (**C**) Lateral profile at $Z = 30$ mm with steering from 0° to 2° to 14°.

The first observation is that the acoustic field consistently exhibits a ripple of approximately 2 dB, attributed to the configuration of empty columns required by the connection system. Secondly, the acoustic pressure remains stable along the aperture projection. When analyzing the steering results, increasing the steering angle leads to a reduction in field amplitude compared to the 0° case. Additionally, a lobe platform emerges at approximately −12 dB, which reduces the dynamic range and introduces out-of-plane reflector signals in that region. For instance, at a steering angle of 10°, a 6 dB loss in the main lobe is observed. In such a scenario, a target located outside the wavefront would produce a reflection only 6 dB below, further contributing to the presence of undesired signals in that area.

Therefore, it can be concluded that even in the case of plane wave imaging, the beam-steering capability is notably constrained. The ability to generate angular diversity is limited to a narrow range between −4° and +4°, resulting in a reduced overlap region within approximately −2 mm to +2 mm. This reduction corresponds to nearly 50% of the insonified area, significantly impacting the effective imaging zone. Additionally, this limitation is accompanied by a 6 dB decrease in dynamic range. Under these conditions, employing plane wave imaging to exploit angular diversity does not appear to be a recommended approach.

3.2. Aperture Performance

Another challenge of operating in the near field with large elements is their limited beam opening angle (approximately 19° in this case), which prevents uniform perception of the insonified region across all elements. As a result, some elements may fail to reach certain areas within the region of interest. This issue has a significant impact on the dynamic range. The beamforming process, which assumes that all elements contribute equally to the focal point, yields a theoretical dynamic range of $20\log(1/\sqrt{N_e})$, where N_e is the number of elements in the square aperture. In this case, this value corresponds to 30 dB. However, since the element contributions are not uniform, the dynamic range does not reach this theoretical value. In this context, each element contributes to the dynamic range depending on the relative position of the reflector with respect to the array, and this contribution can be approximated as follows:

$$DR \approx 20 \log \left(\frac{\max(P_i(\vec{x}))}{\sqrt{\sum_i^{N_e} P_i(\vec{x})}} \right) \quad (1)$$

where $P_i(\vec{x})$ is the reflection detected on the ith-element from the position of the reflector $\vec{x}$. For effective focusing and optimal aperture utilization, it is desirable that the intensity contribution from each element at a given spatial location be of the same order of magnitude. The individual contribution of each element to the image can be assessed by analyzing the intensity with which it perceives each point within the region of interest.

In this study, for each point within the region of interest, we computed the influence applied by the 1024 elements and analyzed how these values are distributed relative to their mean. As an example, we present the distribution of these values, normalized to their mean, at three specific axial positions: Z = 10 mm (Figure 4A), Z = 20 mm (Figure 4B), and Z = 30 mm (Figure 4C). At shallow depths, the differences are notably significant, compromising both the efficient use of electronic resources for certain elements and the overall beamforming efficiency. As depth increases, the intensity distribution becomes more uniform, enhancing the operational balance of the imaging system.

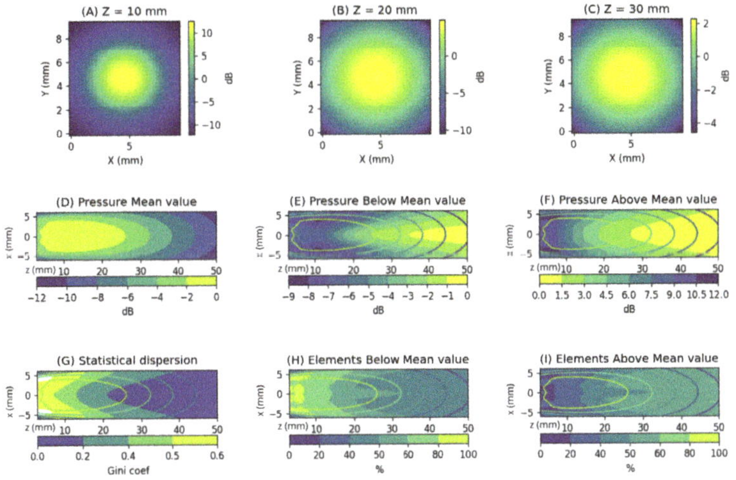

Figure 4. Analysis of element contribution within the region of interest using the full aperture. We considered the region Z = 3–50 mm and X = −5.5–5.5 mm. (**A–C**) Show the amplitude with which the aperture elements influence three points along the axis $(X = 0, Y = 0)$ at depths of (**A**) $Z = 10$ mm, (**B**) $Z = 20$ mm, and (**C**) $Z = 30$ mm. Taking the mean value as a reference, we observe differences ranging from up to 40 dB in (**A**) to around 6 dB in (**C**). (**D**) Shows the average gain level applied at each point in the plane. The values in (**E**) represent the level reached by elements below the mean, while (**F**) shows the values above it. In (**G**), we present the statistical dispersion of these values, calculated using the Gini coefficient. (**H**) Displays the percentage of elements below the mean, and (**I**) shows the percentage above the mean.

Figure 4D shows the average intensity level across the elements, highlighting that maximum pressure is produced between $Z = 10$ mm and $Z = 20$ mm. These curves will be included as reference overlays in all spatial plots. Figure 4E,F illustrate the distribution of values below and above the mean, respectively. Notably, in the region of maximum average intensity, differences of up to 32 dB between extreme values can be observed.

Additionally, we evaluated the percentage of the aperture that contributes above and below the mean intensity level (Figure 4H and Figure 4I, respectively). The requirement for a balanced aperture allowed operation from depths starting at 15 mm. However, as observed in Figure 4E,F, this condition may still result in intensity differences of up to 22 dB, introducing a natural apodization effect across the aperture. These figures indicate that the region of peak pressure is predominantly shaped by the contribution of a limited number of elements.

Finally, to assess the degree of similarity among the elements, we used the Gini coefficient [23,24]. This metric quantifies the level of inequality within a set of values and is particularly useful for analyzing distributions with significant internal variation, as is the case here. Values close to zero indicate high similarity, while values near one reflect high inequality (Figure 4G).

$$G = \frac{\sum_{i=1}^{n}\sum_{j=1}^{n}|x_i - x_j|}{2n^2\,\bar{x}} \quad (2)$$

Values below 0.2 indicate low inequality, values between 0.2 and 0.3 indicate moderate inequality, and values above 0.3 imply high inequality.

The Gini plot shows that below 15 mm, the acoustic field is mainly determined by a small number of elements, resulting in low resource utilization. For the full aperture, a compromise solution can be found operating from 30 mm.

From the simulations, we can deduce that the optimal scanning range capable of covering the full aperture can be approximated as a cylinder from 30 mm to 40 mm with a diameter of 4 mm. In this case, the pulse–echo attenuation is approximately 25.8 dB.

The full aperture requires 1024 electronic channels for operation. Outside of highly specialized laboratories, no system currently supports fully integrated control of this configuration. However, commercial solutions allow partial operation by synchronizing up to four 256-channel systems.

An alternative approach involves operating the aperture with a single 256-channel system combined with a column-wise multiplexer, typically activating eight consecutive columns. This configuration is the most commonly reported in the literature, although detailed operational descriptions are limited. Some studies describe dividing the aperture into seven overlapping windows (50% overlap), each capturing blocks of 32×8 elements. Each acquisition is processed independently and later assembled into a unified 3D image.

Structural imaging in preclinical near-field applications generally yields low-quality results, as many signals contribute minimally to the beamforming process, confirmed by our simulations (see Figure 4). Only through advanced temporal image-differencing techniques can the cerebrovascular system be effectively reconstructed.

4. Sparse Aperture

After describing the challenges of angular plane wave emissions, defining the operating conditions for the full aperture, and outlining the limitations of the hardware system, we propose a solution to these issues based on sparse apertures. A sparse aperture can accommodate hardware constraints, and element selection can be optimized to equalize the overall contribution across the array. Additionally, by using multiple sparse apertures in transmission, we can introduce the necessary diversity into the beampattern to enable effective plane wave imaging without steering. This approach helps reduce artifacts typically associated with conventional plane wave methods, ensures efficient use of electronic resources, and improves beamforming performance. It also aims to uniformly cover the entire region of interest with balanced intensity using a single emission angle, 0°, thereby simplifying the transmission strategy while maintaining image quality.

The objective is not to identify a single optimal aperture, but to define a set of randomly generated apertures within a common framework that exhibit similar main lobe beam pattern characteristics while differing in their sidelobe distributions. This is achieved by keeping the receive aperture constant across transmissions and combining signals directly at the reception channels.

For this study, the aperture is divided into 4×4 element bins arranged on an 8×8 grid. Various apodization modes are applied based on the number of active elements per bin. Within each bin, active elements are randomly selected, introducing diversity while ensuring consistent global behavior across the aperture [9]. Figure 5 shows three representative strategies used in this work.

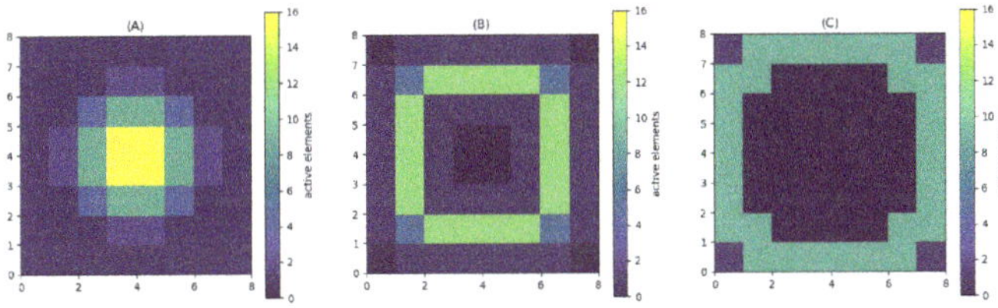

Figure 5. Three sparsification strategies are shown, where the color indicates the number of active elements in each bin. The maximum number of active elements per bin is 16. (**A**) Taylor-based apodization scheme (for emission and reception), (**B**) ring-based apodization scheme (complementary to Taylor apodization in emission), and (**C**) ring-shaped apodization scheme used in reception.

The goal of the transmission is to achieve uniform insonification within the region of interest while enabling rapid energy dispersion outside it, thereby minimizing the impact of external noise sources. The wavefront should remain planar and temporally compact and span the widest possible coverage area, free from secondary fronts that could degrade dynamic range.

To optimize aperture performance, we propose dividing it into two zones: an inner and an outer region. Within each zone, the objective is to homogenize the contribution of active elements while spatially segmenting the insonified area, thus reducing mutual interference between regions.

4.1. Sparse Emission Aperture: Inner Region

We aim to design an aperture capable of generating a wide, flat beam with low sidelobe levels, effectively insonifying the central projection of the aperture. This can be achieved through aggressive windowing, which suppresses the contribution of peripheral elements and enhances main lobe uniformity.

This window must be adapted to the available resources [10,20]. In addition to the spatial discretization already employed (a matrix grid with spacing of $3\lambda \times 3\lambda$), we consider the limitation that the emission amplitude cannot be controlled and is assumed to be fixed for all elements. Consequently, the windowing effect is achieved by adjusting the density of active elements within each bin (see Figure 5).

Ultimately, the resulting aperture should be regarded not as an exact realization of the theoretical model, but as a practical approximation. In Figure 6, we show an example comparing the desired apodization (a Taylor window) with the one actually achieved. The resulting shape is significantly distorted due to the grid structure, which favors alignment in the projection of the elements. The contrast in the obtained field is 10 times lower than desired, although both patterns match along the axis within a range of $\pm 5^{\circ}$ (see Figure 6B).

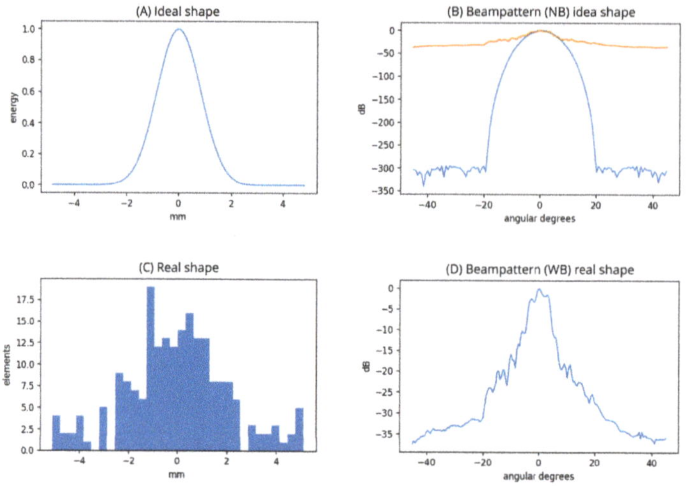

Figure 6. Difference between the desired acoustic field for the apodized transmitting aperture and the field that can actually be achieved given the technical limitations. (**A**) The ideal apodization shape. (**C**) The apodization obtained, represented over the equivalent linear array at axis X = 0, achieved with a binarized aperture, where elements within each bin are randomly selected. (**B**) Displays both the desired field (blue line) and the achieved field (orange line). (**D**) Shows the resulting field obtained through sparsity-based reconstruction.

An aperture designed with a Taylor window was analyzed (see Figure 7), and the contribution of its elements to insonifying the region of interest was evaluated. Balanced participation occurs at depths beyond 10 mm. Although edge and outer elements show lower participation, especially near the aperture, the number of channels with above-average intensity exceeds those with low intensity. This behavior remains consistent with the intended apodization. From 20 mm onwards, the difference between the elements forming the aperture is below 3 dB.

If we observe the acoustic field generated between $Z = 3$ mm and 40 mm in another particular case of this strategy, we see that it has an irregular structure (see Figure 8). However, it drops rapidly—by nearly 25 dB—for X values close to ± 2.2 mm. The distribution of secondary lobes is highly irregular, and as in the case of the full aperture, deflection severely penalizes the dynamic range. It is not possible to go beyond 4°, preventing efficient overlap between deflected images.

The proposed strategy enables the generation of multiple apertures under a common apodization scheme, resulting in similar main lobe behavior while exhibiting distinct side-lobe distributions. Using this approach, we compare the performance of several apertures in Figure 9. For each configuration, the acoustic field was computed at $Z = 30$ mm, spanning the lateral range from $X = -5$ mm to 5 mm, under flat emission conditions. The corresponding wavefront shapes were extracted, and both the similarities and differences among these profiles are relevant to the intended application.

The usable imaging area is limited to a region of 4×4 mm, where the apertures still show small differences, and the behavior of the tails of the wavefront varies in each case. The goal is to leverage the technical limitations that prevent perfect apodization as a tool to generate diversity.

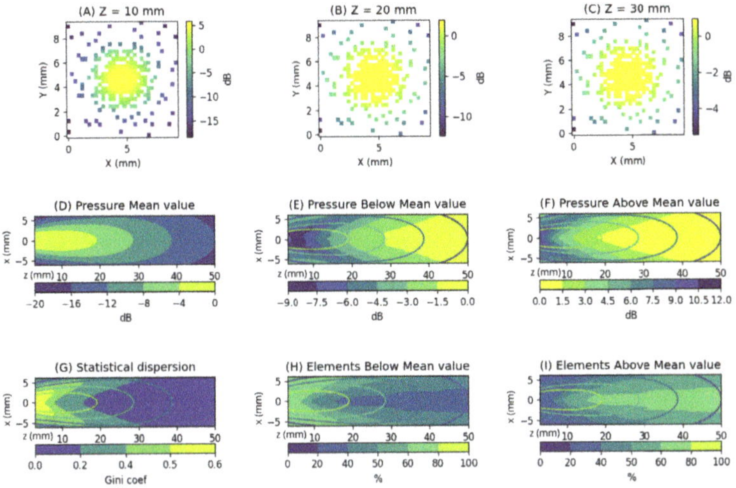

Figure 7. For the Taylor-apodized scheme, analysis of element contributions at different points in space, we considered the region $Z = 3$–$50\,\mathrm{mm}$ and $X = -5.5$–$5.5\,\mathrm{mm}$. (**A–C**) Show the amplitude with which the aperture elements influence three points along the axis $(X = 0, Y = 0)$ at depths of (**A**) $Z = 10\,\mathrm{mm}$, (**B**) $Z = 20\,\mathrm{mm}$, and (**C**) $Z = 30\,\mathrm{mm}$. Taking the mean value as a reference, we observe differences ranging from up to 20 dB in (**A**) to around 6 dB in (**C**). (**D**) Shows the average gain level exerted at each point in the plane. The values in (**E**) represent the level reached by elements below the mean, while (**F**) shows the values above it. In (**G**), we present the statistical dispersion of these values, calculated using the Gini coefficient. (**H**) Displays the percentage of elements below the mean, and (**I**) shows the percentage above the mean.

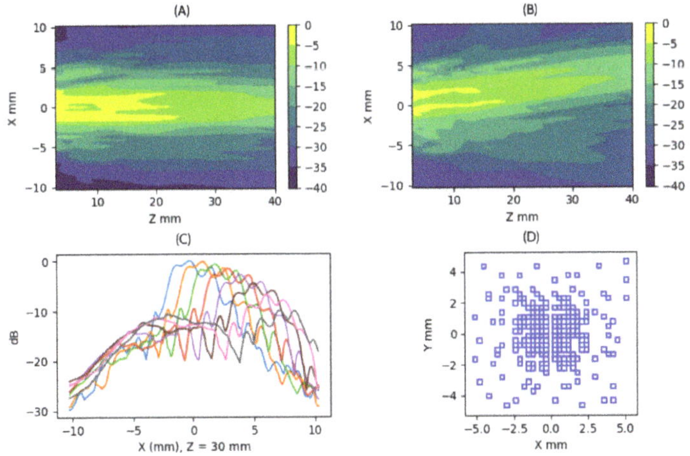

Figure 8. For a particular realization of the Taylor-apodized aperture shown in (**D**), (**A**) displays the field generated by a plane wave emission along the axis $(X = 0, Y = 0)$. (**B**) Shows the field generated by a deflected plane wave emission at 4°. (**C**) Presents the lateral cross-section of the field at $Z = 30\,\mathrm{mm}$, across angles from 0° to 14° in steps of 2°.

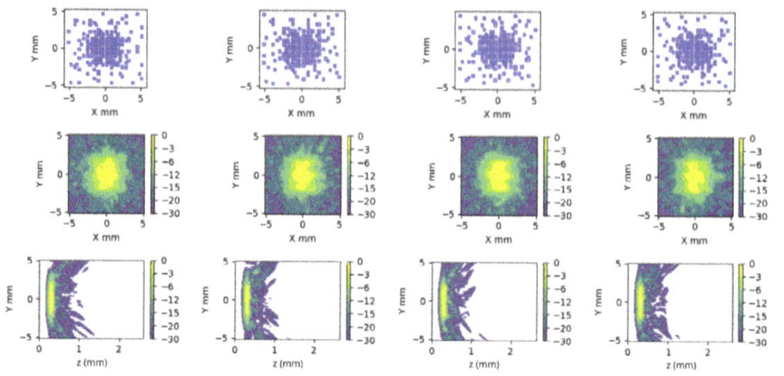

Figure 9. This figure presents four cases of Taylor apodization realizations. In each column, we show the aperture, the generated field at Z = 30 mm within the aperture's projection zone, and the envelope of the wavefront produced by a plane wave emission. The most relevant aspect is the variation observed in the tails of these wavefronts, which are intended to enhance the received signal.

To introduce and exploit signal diversity, the reception aperture must remain fixed. Under this condition, a series of consecutive transmissions can be averaged when received through the same aperture, potentially even directly within the acquisition system. If the transmission aperture remains constant, this averaging primarily reduces electronic noise, thereby improving contrast. However, if the transmission aperture varies between shots, part of the acoustic noise it generates can also be mitigated, further enhancing contrast.

Let the set of signals N_w, corresponding to N_w shots, received in the channel-i be defined as follows:

$$S = \{s_{i1}(t), s_{i2}(t), \ldots, s_{iN_w}(t)\} \tag{3}$$

We can consider different approaches for combining these signals. We define the signal $\hat{s}_i(t)$ as the mean value:

$$\hat{s}_i(t) = \frac{1}{N_w} \sum_{j}^{N_w} s_{ij}(t) \tag{4}$$

We define the signal $\hat{s}_i(t)$ as the one with the minimum absolute value at each time t:

$$\hat{s}_i(t) = \arg \min_{s_{ij}(t) \in S} |s_{ij}(t)| \tag{5}$$

We define $\hat{s}_i(t)$ as the average of the L signals with the smallest absolute values at each time t:

$$\hat{s}_i(t) = \frac{1}{L} \sum_{j \in \mathcal{I}_L(t)} s_{ij}(t) \tag{6}$$

where $\mathcal{I}_L(t) \subset \{1, 2, \ldots, N\}$ is the index set corresponding to the L signals with the smallest $|s_{ij}(t)|$ values at time t.

As an illustrative example, we consider the 1024 received signals corresponding to a single receive element for a scene containing a single reflector, with no added noise, insonified using $N_w = 1024$ different transmit apertures (see Figure 10). For these signals, a histogram of the values at each time instant was computed and visualized as an image, where color indicates the number of coincident values. Over each histogram, the result of combining the signals according to each proposed criterion is shown in red. Below each histogram, the corresponding image of the simulated point is presented, obtained after processing the 256 signals per channel for each data combination strategy.

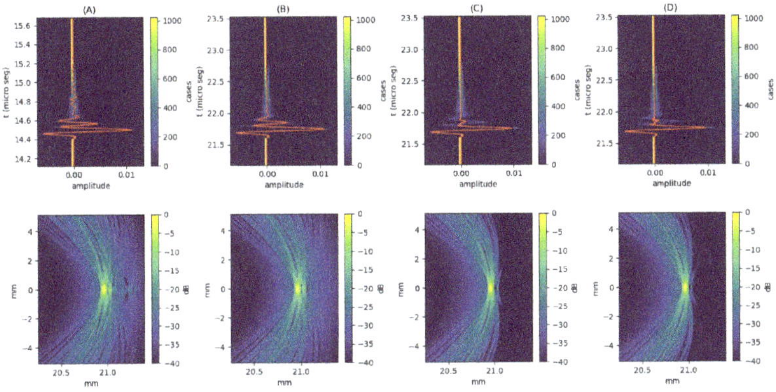

Figure 10. Comparison of different signal combination strategies for a central aperture element. The histogram displays all cases, with a red line indicating the result for each proposed criterion. (**A**) Random acquisition. (**B**) Average of all 1024 acquisitions. (**C**) Average of the 10 lowest-amplitude acquisitions. (**D**) Single lowest-amplitude acquisition.

In Figure 10A, we show a randomly selected case from the 1024 acquisitions, displaying the signal corresponding to a central element of the aperture. In Figure 10B, we present the average of all 1024 acquisitions, again showing the signal from a central element. In Figure 10C, we display the average of the 10 acquisitions with the lowest amplitude among the 1024, focusing on the same central element. Finally, in Figure 10D, we show the individual acquisition with the lowest amplitude, also corresponding to the central element.

What we observe in Figure 10 is that the sequence of signals obtained from multiple transmissions using different apertures enables processing of the raw data prior to beamforming. This processing effectively suppresses acoustic noise generated by the plane wave while preserving the echoes from the main reflectors. When selecting the signals with the lowest amplitudes, we observe a narrowing of the pulse, which leads to an improvement in bandwidth. The improvement is more drastic in this case, but it can also cause a loss in peak value. In the resulting image, this translates into a significant reduction in sidelobes, particularly in their spatial extent, and a noticeable suppression of reverberation artifacts associated with plane wave transmission. It is important to highlight that some of these techniques are compatible with current technology.

To globally assess the performance of the different methods, we calculated, for a reflector located at $(0, 0, 21.22)$ mm, the image generated along the Z-axis for various values of $N_w = 8, 32, 64, 128$. For reception the full aperture was used, and 300 cases were evaluated for configuration. For each point along the axis, Figure 11 shows the maximum value (black line), the minimum (green line), and the mean (red line). As a reference, and to visualize the improvement, a single-shot case $N_w = 1$ has also been included. In all cases, an SNR of 20 dB was applied. Additionally, a line at -25 dB from the main peak is used as a reference for measuring axial resolution (Δz, see Table 1).

Finally, on the curve, in addition to the main peak, two additional points corresponding to other relevant positions on the curve are indicated: $Z = 21.22\ (\text{main}), 21.27, 21.35$ mm. For these points, we calculated their variance and the amplitude difference with respect to the main echo value (see Table 1).

For all cases, the main peak exhibits an axial resolution close to $\lambda = 100$ µm, with very low variance in amplitude. The primary differences among the various results are observed in the secondary peaks that follow the main one. Averaging, regardless of the value of N_w, does not provide a significant improvement in the results. The minimum criterion behaves very aggressively, resulting in a loss in amplitude of approximately 1.6 dB at the reflector position, but it also yields the greatest improvement in dynamic range and can even achieve an axial resolution below λ. However, it shows high variance in the secondary peaks, as a consequence of the inherent randomness of the procedure. The mean of minima method, limited to the four lowest values, provides a more balanced result as N_w increases. Good-quality results are obtained from $N_w = 64$ onwards; from $N_w = 32$, the results offer adequate quality with a reasonable number of shots. Below this value, for high-speed imaging, the minimum criterion appears to be more suitable.

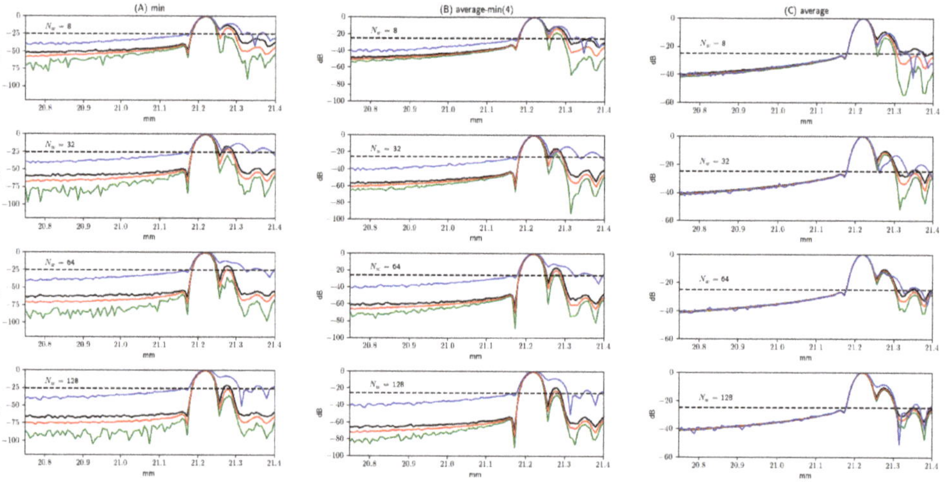

Figure 11. Z-axis image of a reflector located at $(0, 0, 21.22)$ mm obtained for $N_w = 8, 32, 64, 128$ using full reception. For each configuration, 300 images were generated by applying the three proposed signal composition methods: (**A**) minimum, (**B**) average of the four lowest values, and (**C**) full average. For each point along the axis, the maximum (black line), minimum (green line), and mean (red line) values are shown. A specific case for $N_w = 1$ is included in blue for reference. In all cases, a signal-to-noise ratio (SNR) of 20 dB was applied. A line at -25 dB from the main peak is plotted as a reference for estimating the axial resolution (Δz). The blue line represents an arbitrary case.

Table 1. Data extracted from Figure 11 used to evaluate the performance of each configuration (dynamic range at three points, variance, and axial resolution).

N_w	Method	21.22 mm	21.27 mm	21.35 mm	21.22 mm	21.27 mm	21.35 mm	Δz (μm)
			Mean (dB)			Std (dB)		
8	min	0.71	16.47	43.9	0.032	3.05	5.72	100
	average-min(4)	0.34	13.58	34.69	0.018	1.5	4.99	110
	average	0.0	11.37	26.67	0.0	0.91	3.55	123
32	min	1.0	20.46	53.47	0.026	2.85	2.88	97
	average-min(4)	0.75	17.23	45.29	0.019	1.41	2.37	104
	average	0.0	11.35	25.87	0.0	0.38	1.11	123
64	min	1.12	23.37	59.28	0.023	3.64	2.6	97
	average-min(4)	0.9	19.43	51.62	0.015	1.62	1.67	103
	average	0.0	11.34	26.39	0.0	0.37	0.9	123
128	min	1.69	26.31	64.75	0.021	3.75	3.12	58
	average-min(4)	1.49	21.99	57.58	0.016	2.01	1.54	97
	average	0.47	11.37	26.26	0.0	0.25	0.67	123

4.2. Sparse Emission Aperture: Outer Region

The scanning area can be extended by employing a ring-shaped emission configuration. Since we are operating in the very near field, this approach enables sonification of regions beyond the lateral range of ±2.5 mm, complementing the coverage provided by the Taylor emission.

To implement this, the aperture is structured into concentric rings derived from bin segmentation. The dispersion strategy must be carefully designed to suppress edge effects and must account for the distortion caused by central elements, which can radiate several decibels more than the outer elements (see Figure 4). These factors can lead to elevated secondary lobe levels. At the same time, the strategy should ensure coverage over a broad spatial domain to promote diversity in field distributions. Accordingly, the secondary ring is selected within the bin structure, maintaining one element per bin in both the outermost ring and the immediately adjacent inner ring (see Figure 5).

This configuration is analyzed in Figure 12. The ring-based arrangement ensures uniform participation across array elements. However it emits lower energy than the Taylor apodization. The average participation distribution spans a wide area, including the projected footprint of the elements. The Gini coefficient rapidly converges toward an equilibrium state, indicating that element participation becomes balanced from a radial distance of 15 mm onwards. It is observed that, despite the uniform weighting, this configuration involves a portion of the aperture in which the maximum element contribution is 5 dB lower than in the Taylor configuration (see Figure 7).

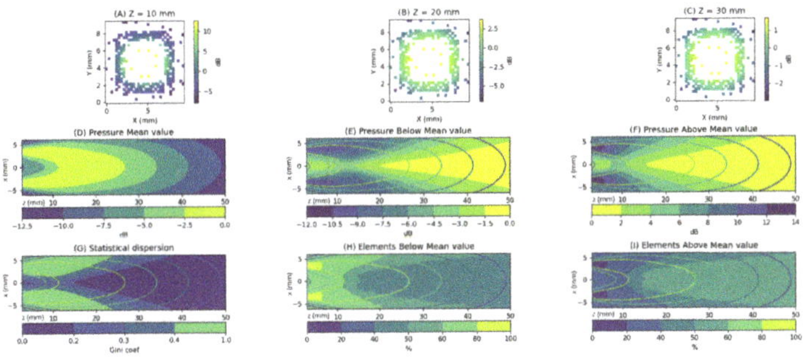

Figure 12. Analysis of the ring-shaped apodized aperture at different points in space; we considered the region $Z = 3$–50 mm and $X = -5.5$–5.5 mm. (**A–C**) Show the amplitude with which the aperture elements influence three points along the axis ($X = 0, Y = 0$) at depths of (**A**) $Z = 10$ mm, (**B**) $Z = 20$ mm, and (**C**) $Z = 30$ mm. Taking the mean value as a reference, we observe differences ranging from up to 15 dB in (**A**) to around 4 dB in (**C**). (**D**) Shows the average gain level exerted at each point in the plane. The values in (**E**) represent the level reached by elements below the mean, while (**F**) shows the values above it. In (**G**), we present the statistical dispersion of these values, calculated using the Gini coefficient. (**H**) Displays the percentage of elements below the mean, and (**I**) shows the percentage above the mean.

Based on this strategy, four aperture configurations were designed, and the resulting fields and corresponding planar wavefronts were computed (Figure 13). All configurations exhibit a planar wavefront formed within the region between 2.5 mm and 5 mm, leaving the central zone (±2.5 mm) with a decay margin ranging from 6 dB to 12 dB and a disordered distribution. This behavior helps reduce the presence of imaging artifacts originating from the central region.

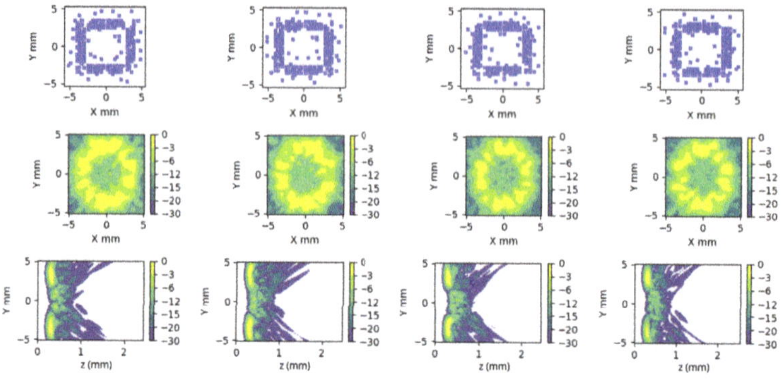

Figure 13. This figure presents four cases of ring-shaped apodization realizations. In each column, we show the aperture, the generated field at Z = 30 mm within the aperture's projection zone, and the envelope of the wavefront produced by a plane wave emission.

For demonstration purposes, we consider another example with a different number of emission apertures. Using a line of reflectors distributed along the Z-axis every 250 μm as a reference, we simulated $N_w = 128, 64, 32$ and observed (see Figure 14) the acquisition at a specific element located at position $(-4.2, -2.25)$. For these signals, knowing that they all correspond to the same scene and share common elements, we applied three different processing methods to reduce noise.

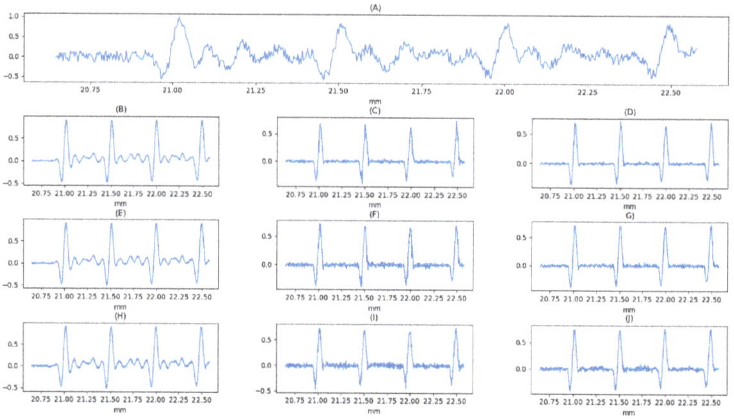

Figure 14. The figure shows the signal received by an element located at position $(-4.2, -2.25)$, corresponding to a plane wave emission performed using a series of random apertures defined according to a ring-shaped apodization. The simulation includes attenuation caused by coupling, and 20 dB of noise has been added relative to the echo with the highest gain. In (**A**), the signal obtained from a single shot is shown. (**B–D**) Correspond to 128 shots and represent, respectively, the mean, the minimum, and the mean of the four lowest values. (**E–G**) Correspond to 64 shots and likewise represent the mean, the minimum, and the mean of the four lowest values. (**H–J**) Correspond to 32 shots and again represent the mean, the minimum, and the mean of the four lowest values.

The first method, averaging, reduces electronic noise but is insufficient to eliminate all secondary oscillations characteristic of a flat wavefront (B, E, H). The second method, minimum selection, is a more aggressive approach that yields high contrast without secondary lobes; however, it is highly sensitive to the random configuration of apertures and introduces high-frequency noise typical of nonlinear systems (C, F, I). The third method, averaging a subset of the lowest-amplitude acquisitions, provides a smoother response with a significant reduction in secondary lobes, though not as pronounced as with minimum selection. The number of acquisitions used in the subset depends on the desired level of smoothness; in this case, we averaged the four lowest values (D, G, J).

The original signal is shown in Figure 14A. In the simulation, additive noise was introduced at a level of 20 dB relative to the maximum peak amplitude across all acquisitions. Figure 14B–D correspond to 128 emissions; Figure 14E–G to 64 emissions; and Figure 14H–J to 32 emissions. Disregarding the arithmetic mean—which does not yield a significant improvement—the most favorable results are obtained using either the minimum value or the mean of the minimums. Clearly, as the number of acquisitions increases, the likelihood of achieving a more accurate reconstruction also improves. When the acquisition count is high, the mean of the minimums produces highly satisfactory results. However, under constraints on the number of shots, using the minimum value in combination with appropriate filtering techniques offers a robust and efficient compromise.

Finally, Figure 15A,B show the diffraction pattern of the two emission apertures within the plane of interest. To assess how they complement each other, a combined representation of the mean field from both configurations is also included (see Figure 15C). The Taylor emission is approximately 6 dB stronger than that of the ring. It should also be noted that the insonification strategy for the outer region is 10 dB lower than that of the inner region, and similarly, the reflectors in the outer zone are 20 dB weaker when insonifying the inner region. This creates a degree of isolation between the two zones, which can help reduce interference in complex scenes.

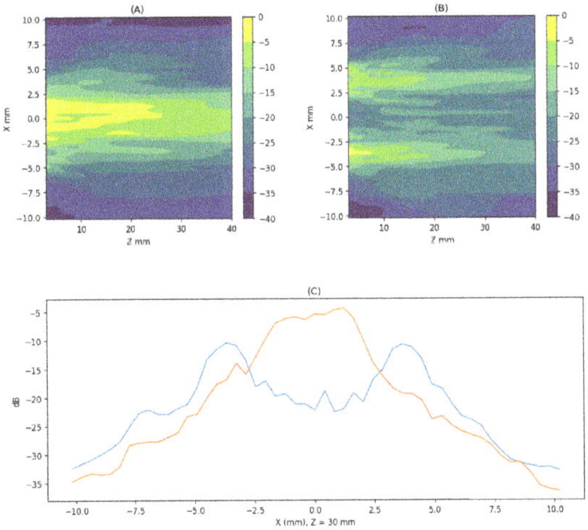

Figure 15. The figure shows how the two emission modes complement each other in a plane wave emission with a deflection angle of 0°. (**A**) Displays the field generated by a Taylor apodization. (**B**) Shows the field generated by a ring-shaped aperture. (**C**) Presents the average of both fields. Although they complement each other well, a drop of approximately 10 dB is observed between the Taylor apodization (original line) and the ring-shaped apodization (blue line).

One aspect to consider is that, if the same reception aperture is used for both emissions, it is possible to combine the results from both acquisitions on the same signals:

$$\hat{s}_i(t) = \hat{s}_{iR}(t) + K\,\hat{s}_{iT}(t) \tag{7}$$

where $\hat{s}_{iT}(t)$ is the signal processed from the Taylor-apodized emission, $\hat{s}_{iR}(t)$ is the signal processed from the ring-shaped emission, and K is a weighting factor, empirically determined in this case as $K = 0.4$.

In Figure 16, we can see how this process works. In this case, the full aperture was used for reception. First, for a value of $Z = 20.9$ mm, the curve defining the viewing area was calculated (blue line in Figure 16A). The combination of both excitations allows coverage of a region of ± 4.5 mm with a ripple of approximately 1 dB and a drop of 9 dB at 5 mm from the center of the aperture. To test how the regions perform, four targets were placed at the same depth at positions $X = \{-6, -4, 0, 6\}$. The outermost values fall outside the field of view, while the other two belong, respectively, to the ring and the inner zone. Figure 16B shows the beamforming results using $\hat{s}_{iR}(t)$ (orange line), $\hat{s}_{iT}(t)$ (blue line), and the combined signal $\hat{s}_i(t)$ (green line); see also Figure 16A. All signals were generated using the criterion of the average of the four lowest-amplitude acquisitions over 64 emissions. Finally, Figure 16C shows the resulting image.

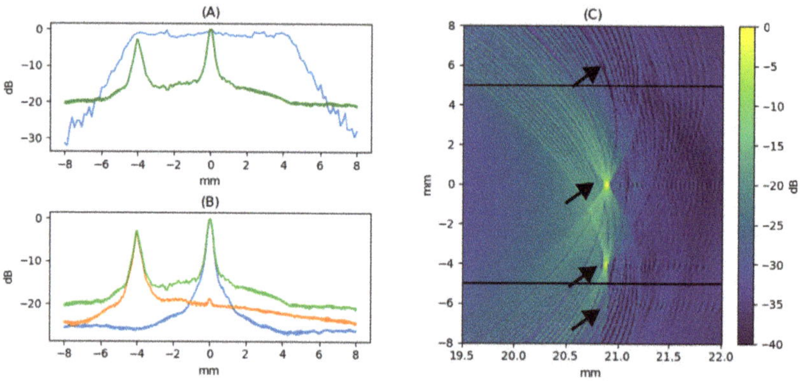

Figure 16. Visualization. of the combined insonification strategy. (**A**) Viewing area curve at $Z = 20.9$ mm, showing arraycoverage of ± 4.5 mm with a 1 dB ripple and 9 dB drop at 5 mm from the center (blue line). (**B**) Based on the average of the four lowest-amplitude acquisitions over 128 emissions, beamforming results using ring-apodized signal $\hat{s}_{iR}(t)$ (red), Taylor-apodized signal $\hat{s}_{iT}(t)$ (blue), and combined signal $\hat{s}_i(t)$ (green, also in (**A**)). (**C**) Final reconstructed image showing target visibility across regions. The positions of the reflectors are indicated with arrows.

One of the most relevant outcomes is the effective separation between the inner and outer regions, which minimizes cross-interference and reduces the influence of elements outside the imaging area, including grating lobes generated from outside the field. The combination of signals $\hat{s}_{iT}(t)$ and $\hat{s}_{iR}(t)$ enables reconstruction of the complete image. In this sense, everything in Figure 16C that lies beyond the marked boundaries consists of lobes caused by reflectors within the region of interest and therefore should not be considered part of the image.

From a system design perspective, the ability to construct $\hat{s}_i(t)$ directly within the acquisition system contributes to improved signal acquisition and compression capabilities. However, it can also reduce the dynamic range and should therefore be implemented with appropriate control. If each emission is processed independently, it has the advantage of reducing cross-interference between the two regions.

The use of multiple transmit apertures during reception, in conjunction with non-linear processing operators such as the minimum function, facilitates the acquisition of echo signals with enhanced bandwidth. As a consequence, axial resolution is improved, approaching the order of one wavelength. Additionally, by inherently suppressing grating lobes, this methodology simplifies the design constraints of the receive aperture and enhances overall image fidelity.

4.3. Sparse Reception Aperture

The proposed processing approach for plane wave imaging contributes to enhancing both dynamic range and axial resolution. Additionally, by centering the imaging region over the aperture projection, it helps mitigate external interference. However, lateral resolution is primarily determined by the receive aperture. The increased bandwidth, combined with the narrow steering angle within the region of interest, ensures that grating lobes do not constitute a design constraint.

Design choices must meet the requirements of providing optimal resolution and enhancing contrast. Under ideal conditions, an isolated target imaged with a full 32×32 narrowband aperture would theoretically yield a dynamic range of approximately -30 dB. For a sparse aperture with 256 elements, the theoric dynamic range varies between -30 and -18 dB depending on the spatial distribution of the elements. If the design is not appropriate, and considering that the system operates in the very near field, these values may be even lower (see Equation (1)). In scenarios with multiple targets, the interaction of various secondary wavefront patterns reduces these margins.

It becomes feasible to reconstruct the full aperture in reception using four acquisitions of 256 elements (see Figure 17(A-1–A-4)). For this reception configuration, we have simulated the image generated by the full aperture for a series of targets distributed along the axis Z $(X = 0, Y = 0)$ and present a detailed view. Additionally, we show the signal received at the element located at (0.3, –0.15) for this image. Two emission cases have been simulated: (B) the full aperture and (C) the Taylor emission-based dispersion strategy ($N_w = 8$, min-method). As observed, in the case of full emission, the secondary wavefronts from the plane wave introduce oscillations that elevate the background amplitude, thereby reducing the dynamic range. In Figure 17C, the dispersion strategy achieves a higher dynamic range. In Figure 17D, we present the signal received at the element located at (0.3, –0.15) for both emissions. Although our approach incurs some loss in reflectivity in certain targets, the resulting image exhibits an increased dynamic range and reveals the underlying interference structure between the secondary lobes of each reflector.

At this stage, the design of the receive aperture becomes a more classical problem, primarily focused on ensuring both contrast and lateral resolution. This is constrained by technical factors such as the number of elements and their sensitivity, as illustrated in Figure 4. In this context, it remains a priority to maintain balanced energy across the receive channels. The Taylor and ring-shaped aperture designs proposed for transmission meet this requirement but exhibit two very different operational patterns: the Taylor aperture achieves low sidelobes and low resolution, while the ring aperture results in high sidelobes and high resolution. In Figure 18, we repeat—under the same conditions as in Figure 17—the simulation for reception based on Taylor apodization and ring apodization. Additionally, we compare these with the full aperture reconstruction and the conventional plane wave approach from Figure 17, evaluating the lateral resolution Δx at -6 dB over the target at $(0, 0, 20.5)$ achieved by each configuration.

In Figure 18A,C, we show an aperture following Taylor apodization and its corresponding generated image. This configuration emphasizes the suppression of sidelobes but results in poor lateral resolution. In Figure 18B,D, we present an aperture arranged in a ring distribution, selected to enhance lateral resolution, along with its corresponding image. This setup improves resolution but exhibits elevated sidelobes, which reduce the dynamic range.

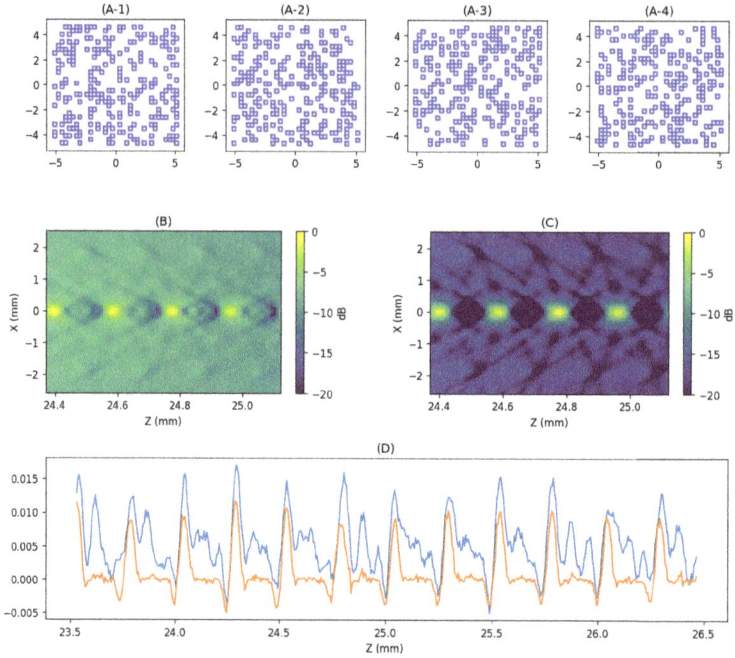

Figure 17. For a scenario in which a sequence of reflectors spaced 250 µm apart has been placed, (**B**) was computed using a full aperture from a flat emission. A composite image was also generated from four acquisitions (**A-1–A-4**), each consisting of eight emissions based on Taylor apodization and processed using minimum absolute techniques ($N_w = 8$, min-method) (**C**). (**D**) Shows a detailed view of the signal obtained at the element located at (0.3, −0.15). The signal corresponding to the full aperture emission is shown in blue, while the one corresponding to the Taylor-apodized and minimum-processed emission is shown in orange.

If we examine the lateral profiles in Figure 18E, ranked from worst to best, we have conventional plane wave (FULL), $\Delta x = 500$ µm, with a dynamic range of −7.4 dB; Taylor (TAYLOR), $\Delta x = 500$ µm, with a dynamic range of −7.4 dB; sparse–full (SP_FULL), $\Delta x = 326$ µm, with a dynamic range of −16.4 dB; and ring (RING), $\Delta x = 214$ µm, with a dynamic range of −17 dB and a sidelobe level of −11 dB.

While for axial resolution we have been able to achieve values on the order of λ or even smaller, the lateral resolution is determined by the behavior of the apodization window. Theoretically, the resolution for this aperture is around 400 µm, which is four times greater than the axial resolution. The values obtained for the different implementations range between 500 µm and 200 µm, with a trade-off associated with increased sidelobe levels. Our conclusion is that, once the challenge of grating lobes has been avoided, the design of the receive aperture should be primarily guided by the application's requirements for contrast and resolution. The most versatile option is to acquire the full aperture and then apply digital signal processing techniques to achieve the desired objectives. However, this solution increases the acquisition time by four.

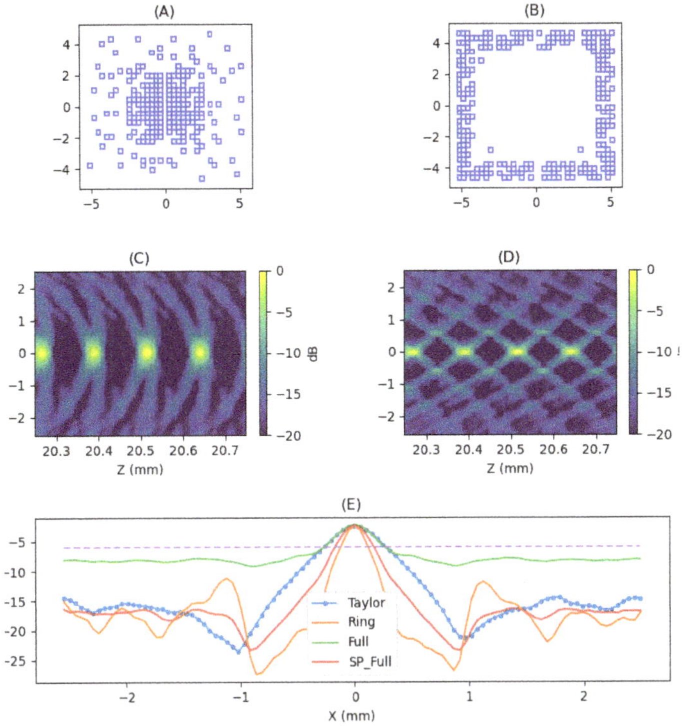

Figure 18. For the same scenario described in Figure 17, the image obtained using a reception aperture based on (**A**,**C**) Taylor apodization and (**B**,**D**) a ring-shaped aperture mounted on the outer bins is presented. (**E**) Displays the lateral profile of the four reception configurations over the target at $(0, 0, 20.5)$. The -6 dB threshold (purple dash line) serves as a criterion for lateral resolution.

As an example, one solution is to compose the image using the geometric mean of the Taylor apodization and the ring apodization. In Figure 19, we show the result of this combination for the same scenario as in Figures 17 and 18. The outcome is similar to that obtained with SP_FULL, featuring a slightly lower dynamic range and slightly higher lateral resolution, but with practically equivalent performance.

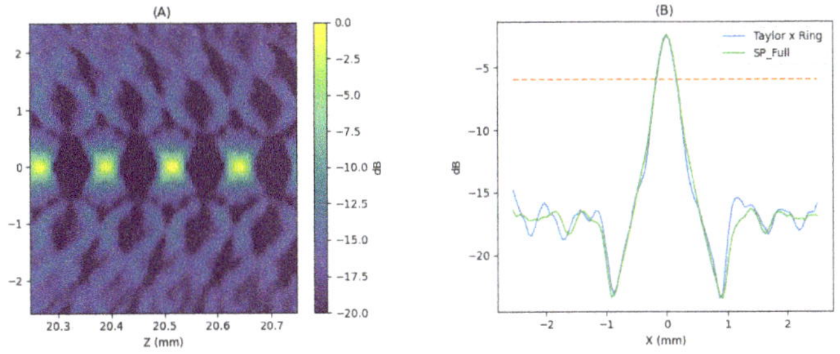

Figure 19. For the same scenario described in Figure 17, (**A**) was obtained using a reception aperture based on the geometric average of the configurations (see Figure 18C,D). (**B**) Displays the lateral profiles of the four reception configurations over the target at $(0, 0, 20.5)$, along with the corresponding profile for SP_FULL for comparison. The -6 dB threshold (orange dash line) serves as a criterion for lateral resolution.

5. Imaging System

The technological implementation of the presented solution involves addressing several key challenges. First, depending on the selected operational mode, the system requires the development of two to three independent multiplexing structures: two for transmission (potentially shared with reception) and one specifically for reception. The first structure uses a Taylor apodization scheme; the second employs a ring-shaped scheme; and the third, dedicated to reception, is designed to cover the full aperture in a maximum of four firings.

Second, since beamforming does not require knowledge of the position of the emission elements, the system could incorporate a mechanism to autonomously and randomly select transmitting apertures within the constraints of each transmission strategy. Additionally, a processing stage could be implemented in the receiving channels to apply a nonlinear, minimum-based filter functioning as an EMI-like filtering stage [25].

A final example has been developed to evaluate a more complex scenario: three double-point curves were designed, spaced 125 μm apart in the Z-axis and 500 μm in the X-axis. This configuration is illustrated in Figure 20, which shows the image obtained using the full aperture. In the image, the targets are visible but exhibit a very low dynamic range, close to 3 dB.

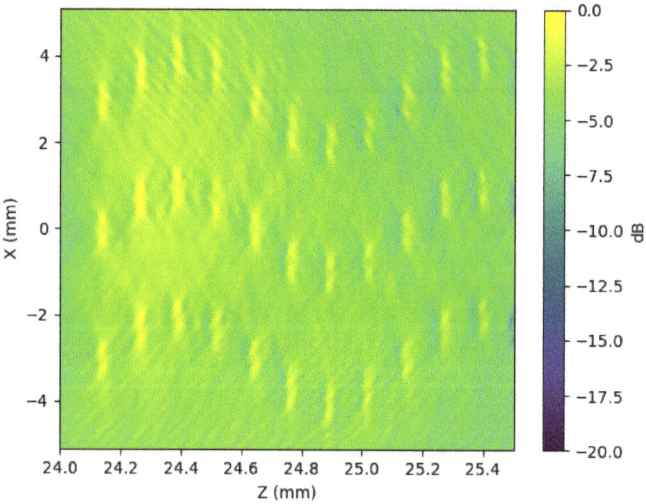

Figure 20. Image generated using plane wave emission with a deflection angle of 0° and full aperture. The scenario consists of three curves formed by pairs of points spaced 125 μm apart in the X direction and 500 μm in the Z direction.

For the same scenario, we computed the image using our double-emission scheme. First, the full aperture reconstructed from the four complementary acquisitions was used, and second, a ring-shaped aperture combined with a Taylor-tapered apodization aperture was used, integrated geometrically after beamforming.

In the first case (see Figure 21), the two regions are presented separately. For the inner region, the image displays only the central part of the scene, which is reconstructed with good resolution and a dynamic range close to 15 dB. In contrast, the image of the outer region, while maintaining resolution, shows a reduction of approximately 50% in dynamic range.

In Figure 21C, both images are combined by merging the data during the beamforming process (which can also be performed in the reception channels). To preserve balance in the final image, the central region was attenuated by 6 dB.

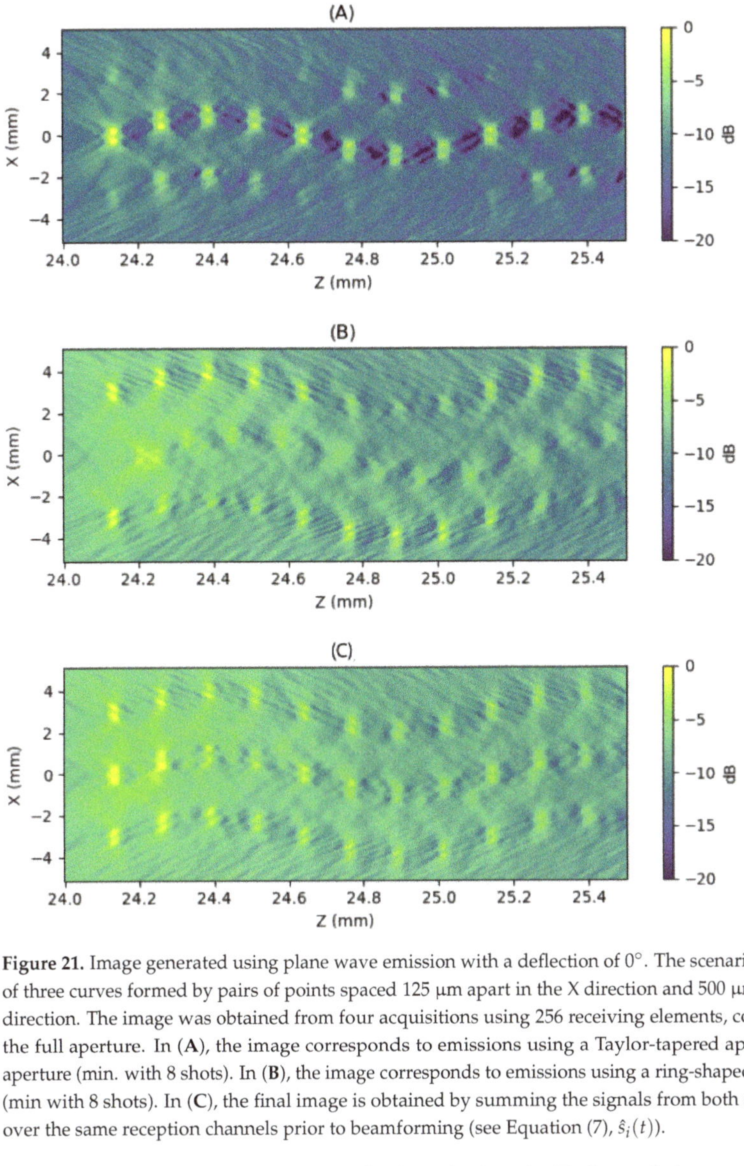

Figure 21. Image generated using plane wave emission with a deflection of 0°. The scenario consists of three curves formed by pairs of points spaced 125 µm apart in the X direction and 500 µm in the Z direction. The image was obtained from four acquisitions using 256 receiving elements, completing the full aperture. In (**A**), the image corresponds to emissions using a Taylor-tapered apodization aperture (min. with 8 shots). In (**B**), the image corresponds to emissions using a ring-shaped aperture (min with 8 shots). In (**C**), the final image is obtained by summing the signals from both emissions over the same reception channels prior to beamforming (see Equation (7), $\hat{s}_i(t)$).

Finally, Figure 22 shows the result of combining the Taylor and ring strategies in reception. In these two cases, the two regions are already presented as combined by Equation (7) (Figure 22A,B), and the following image displays the result of applying the geometric mean between both apertures. The outcome is similar to that presented in Figure 21, with two reception apertures instead of four.

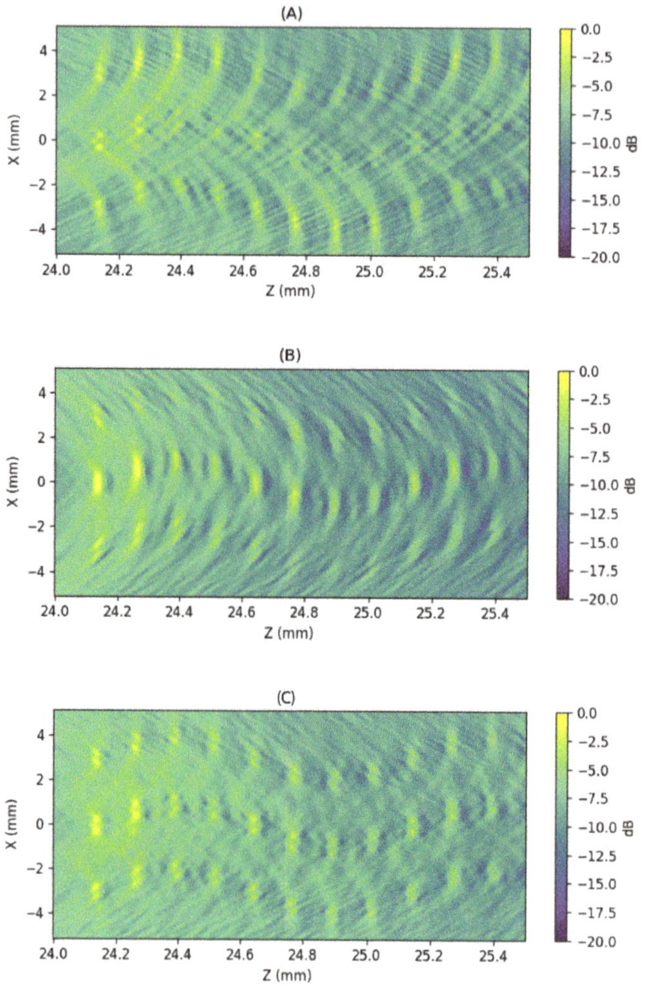

Figure 22. Image generated using plane wave emission with a deflection angle of 0°. The scenario consists of three curves formed by pairs of points spaced 125 μm apart in the X direction and 500 μm in the Z direction. (**A**) Was obtained using 256 receiving elements, modeled with a Taylor apodized aperture, by summing the signals from both emissions (min with 8 shots) over the same reception channels prior to beamforming. (**B**) Was obtained using 256 receiving elements, modeled with a ring-shaped aperture, also by summing the signals from both emissions (min with 8 shots) over the same reception channels prior to beamforming. (**C**) Was obtained by combining (**A**,**B**) using a geometric mean.

System Performance

After analyzing the results and considering the necessary trade-off between application requirements and system performance, we propose that, based on the described operating method, the aperture can be effectively set between 20 mm and 30 mm. In pulse–echo mode, this results in 15 dB of attenuation within the sample and 7.2 dB in the coupling material (a total attenuation of 22.2 dB), compared to the 31 dB caused by increasing the radiating surface. If necessary, we can move the acquisition forward to 15 mm at the expense of reducing the homogeneity of the element contributions within the ring configuration.

Each emission cycle lasts 40 μs, with data acquisition occurring over 13.3 μs. With a sampling rate of 62.5 MHz (14 bits) across 256 channels, each acquisition generates 560,000 bytes. For a communication channel of 3 GB/s, the transfer time is 186.6 μs, while for a Gigabit Ethernet system, it is 2.797 ms. This means that, depending on the communication technology, it is possible to parallelize the transmission of one acquisition with either 3 or up to 74 firings.

Based on these timings and assuming maximum transfer speed, different operating modes can be selected depending on the desired image speed and/or quality. For example, using a minimum-based method with $N_w = 9$ and two acquisitions, the fastest imaging rate is achieved, with 893 volumes per second (approximately $9 \times 9 \times 10$ mm^3). To increase image quality, using a reception based on four subapertures (SP_FULL) and an emphaverage-min criterion with $N_w = 32$, a rate of 62 volumes per second can be obtained. In the case of using the combination of the Taylor aperture with the ring in reception, a rate of 124 volumes per second can be achieved.

Disregarding the computational cost of beamforming, we can state that, depending on the configuration used, the system is capable of providing a spatial resolution of between ($200 \times 200 \times 100$ μm^3) and ($300 \times 300 \times 100$ μm^3) and a temporal resolution ranging from 893 to 62 Hz. In the literature, the only example that provides relevant data with the same aperture reports a spatial resolution of ($200 \times 280 \times 175$ μm^3) and a temporal resolution of 6 Hz [20]. In another work [21], for a 7 MHz (32×32) aperture based on the same four-region multiplexing system, configurations are presented that enable the generation of volumes with 4, 10, and 16 aperture pairs (with five plane waves per pair), achieving temporal resolutions of 750 Hz, 300 Hz, and 187 Hz, respectively.

6. Experimental Results

In order to experimentally verify the feasibility of using spatial diversity from sparse apertures to improve image quality in plane wave imaging as an alternative to beam steering, we employed a 16×16 element array with a center frequency of 5 MHz $\pm 10\%$ (60% bandwidth), manufactured for non-destructive testing (NDT) (Doppler S.L., Guangzhou, China https://www.cndoppler.com/ (accessed on 9 October 2025)). A SITAU 1 system (DASEL S.L, https://www.daselsistemas.com, (accessed on 9 October 2025)) with up to 128 parallel channels multiplexed to 256 was used as the working platform.

The inter-element spacing is 0.9×0.9 mm, with each element measuring 0.7×0.7 mm. Operating in water, this corresponds to an inter-element distance of approximately $3\lambda \times 3\lambda$ ($\lambda = 300$ μm). This aperture is similar to the one we have proposed for our application, although it operates at a lower frequency and contains fewer elements, thus offering reduced spatial diversity. To make the problem equivalent, we propose reducing the number of active elements from 256 to 64. The experimentation is limited to resolving the inner region.

Unfortunately, the 16×16 configuration offers fewer opportunities to exploit spatial diversity compared to the 32×32 aperture, and it does not easily conform to the dispersion schemes proposed in the previous sections. To preserve discretization capability across the aperture, we opted to organize it into 2×2 bins, with a maximum of four active elements per bin. This scheme yields a total of 84 active transducers.

Since it is desirable for the central bins of the aperture to maintain a high level of participation, in order to reduce the number of channels to 64, the remaining bins must undergo a random selection (purging) process. Figure 23 shows (A) the desired apodization strategy, (B) its adaptation to the available aperture and resources, and (C) several resulting dispersion strategies, from which the emission apertures are randomly selected.

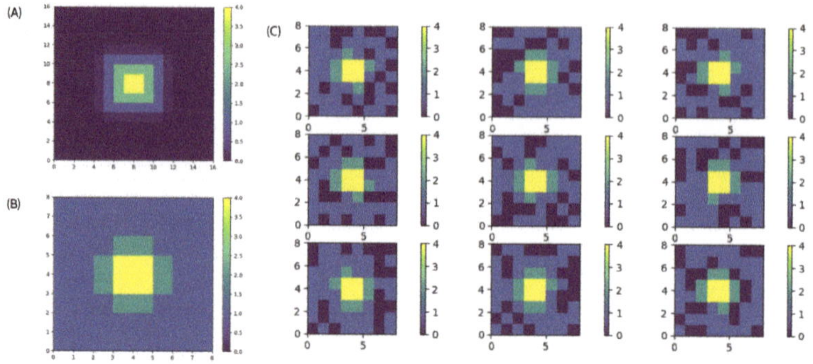

Figure 23. (**A**) The desired apodization strategy for a 16 × 16 array, (**B**) its adaptation as a sparsity strategy to match the available aperture and resources, and (**C**) several resulting dispersion strategies.

For simplicity, reception is performed using the full aperture, constructed from four complementary subapertures. For each experiment, three images were obtained: the *full plane wave* image, generated by transmitting a single plane wave using the entire aperture and receiving with the full aperture, and our proposed solution in two configurations for $N_w = 32$, the minimum criterion (*min*) and the average of the four lowest signals (*average-min(4)*).

A total of four experiments were designed. Figure 24 shows several images of the experimental setup. The experiments conducted include the following: imaging the head of a pin; imaging two filaments at different depths; imaging seven parallel wires at the same depth; and volumetric imaging of a rat skull. Due to the reduced aperture size, the imaging region is limited to $X = [-2, 2]$, $Y = [-2, 2]$ and depths between 10 mm and 20 mm. All the reconstructed images include contour lines at -6 dB and -12 dB, used to assess spatial resolution and energy spread.

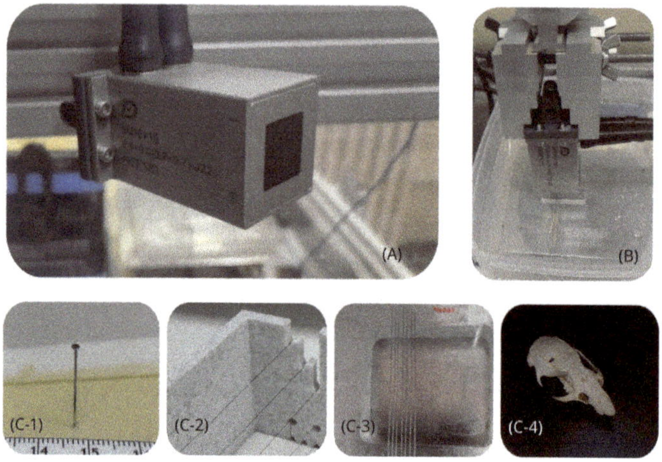

Figure 24. (**A**) The 5 MHz 16 × 16 element 2D array from Doppler; (**B**) the experimental setup; and the four scenarios used in the experiments: (**C-1**) imaging the head of a pin; (**C-2**) two filaments at different depths; (**C-3**) seven parallel wires at the same depth; (**C-4**) and the rat skull.

The first experiment (see Figure 25) is designed to demonstrate the improvements achieved in the received ultrasonic signals. In the top row, the signal from element 106 shows a clear reduction in interference caused by secondary wavefronts. This improvement is consistently observed across all channels. Since the target is the head of a pin, the echo naturally includes contributions from surrounding structures, which remain visible in the final image despite signal processing—especially in channels located near the edges of the aperture. When comparing the reconstructed images, the *average-min(4)* technique yields the highest image quality, combining the benefits of the *min* approach with significantly reduced high-frequency noise. The enhancement is particularly evident in axial resolution and in the suppression of noise along the axial direction.

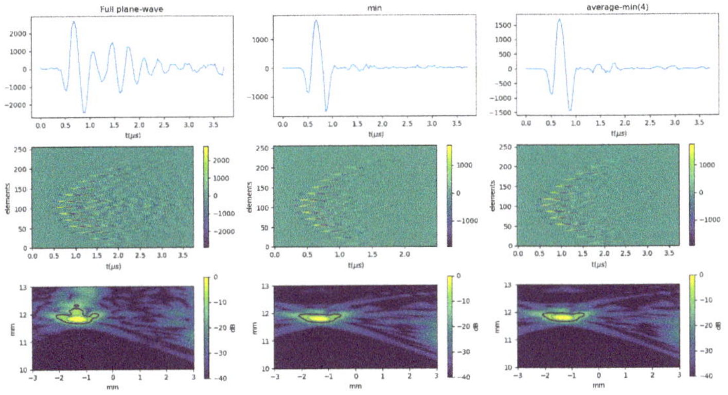

Figure 25. Comparison of three imaging techniques, *full plane wave*, *min*, and *average-min(4)*, applied to a point-like reflector (pinhead). The top row shows the signal from a single channel, the middle row displays the full dataset from 256 elements, and the bottom row presents the reconstructed images.

Table 2 presents the resolution data for Figure 25. The two proposed techniques yield similar results, both outperforming the full plane wave method. However, the pinhead produces a complex echo response, and the entire pulse cannot be attributed solely to plane wave reverberation. If we consider the total pulse length Δz (it is produced at -20 dB), it exceeds the wavelength, and at this depth, a small grating lobe (-20 dB) is generated at 4.1 mm from the target. With the target located at 1.33 mm, this results in a displacement of 2.67 mm, which falls outside the valid imaging zone defined between -2 mm and 2 mm.

Table 2. Resolution data measured in the pinhead experiment.

	$\Delta z_{-6\mathrm{dB}}$	$\Delta z_{-12\mathrm{dB}}$	Δz	$\Delta x_{-6\mathrm{dB}}$	$\Delta x_{-12\mathrm{dB}}$
full plane wave	270 μm	573 μm	1000 μm	800 μm	1600 μm
min	210 μm	300 μm	800 μm	565 μm	1480 μm
average-min(4)	210 μm	300 μm	565 μm	800 μm	1480 μm

The second experiment (see Figure 26) features two wire reflectors positioned at different depths, enabling a direct evaluation of axial resolution and dynamic range. In addition to the reconstructed images, the figure includes maximum intensity projections along the principal axes (depth and lateral), comparing the *average-min(4)* technique (red line) with the full plane wave approach (blue line).

The results demonstrate a consistent improvement in both dynamic range and axial resolution with the proposed method. Specifically, the axial resolution is enhanced ($\Delta z_{-6\,\mathrm{dB}} = 200$ μm, $\Delta z_{-12\,\mathrm{dB}} = 280$ μm), allowing for clearer separation between closely spaced reflectors. In this case, no grating lobes are present; only secondary lobes are produced.

Furthermore, the image processed with our technique exhibits a significant reduction in background noise, which enhances the visibility of secondary lobes. Although these lobes appear more pronounced due to the lower noise floor, they remain well-contained, indicating that the method preserves spatial fidelity without introducing artifacts.

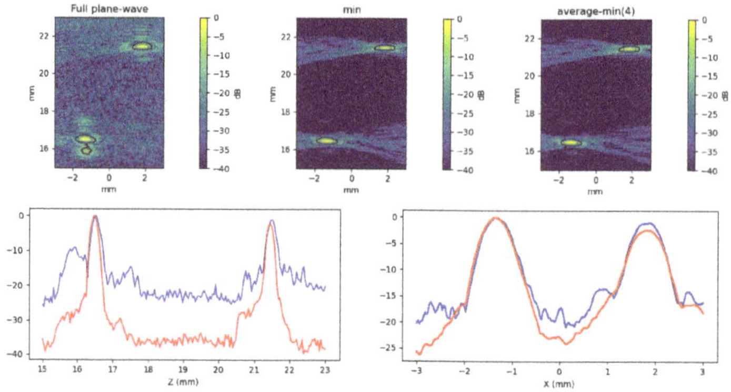

Figure 26. Imaging of two reflectors at different depths using *full plane wave*, *min*, and *average-min(4)* techniques. The top row shows the reconstructed images, while the bottom row presents maximum intensity projections along the axial (Z) and lateral (X) directions. The blue line is the *full plane wave*, and the red line is the *average-min(4)*. Contour lines at −6 dB and −12 dB highlight the improved resolution and energy localization achieved.

The third experiment (see Figure 27) evaluates the imaging of four wires (acupuncture needles, 250 μm diameter) spaced 1 mm apart, positioned at slightly different depths. The *average-min(4)* technique shows the clearest separation within the −6 dB contours and the lowest background noise, while also preserving the geometric arrangement. Compared to *min* and *full plane wave*, it minimizes overlap and secondary lobes, confirming its effectiveness for resolving closely spaced reflectors with axial misalignment. We would like to highlight the difficulty of resolving an image of this type using *full plane wave*, where reverberations are stronger than the actual echoes from the reflectors.

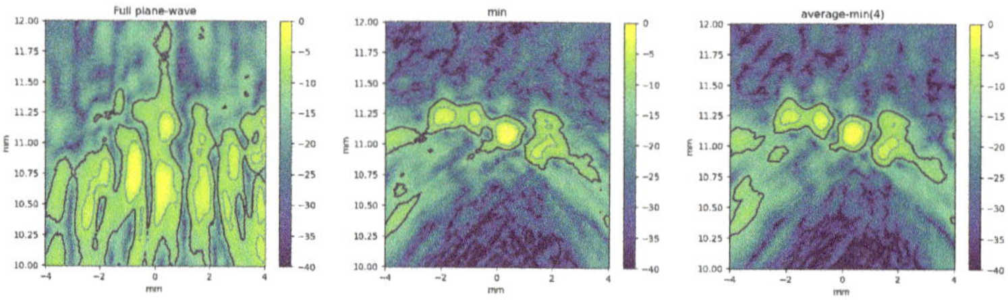

Figure 27. Reconstruction of four wires spaced 1 mm apart, positioned at slightly different depths. The images compare *full plane wave*, *min*, and *average-min(4)* techniques. Contour lines at −6 dB and −12 dB are used to assess lateral resolution and energy spread.

The fourth experiment (see Figure 28) assesses the ability of each technique to image through bone tissue using a rat skull. A reflector placed behind the skull serves to evaluate wave penetration. The figure includes a photograph of the skull and 3D reconstructions using *full plane wave*, *min*, and *average-min(4)*. *Full plane wave* shows poor localization and high dispersion. The *min* technique improves focus and reduces noise. *Average-min(4)* yields the clearest result, with a well-defined reflector and minimal background interference, confirming its superior performance in imaging through bone.

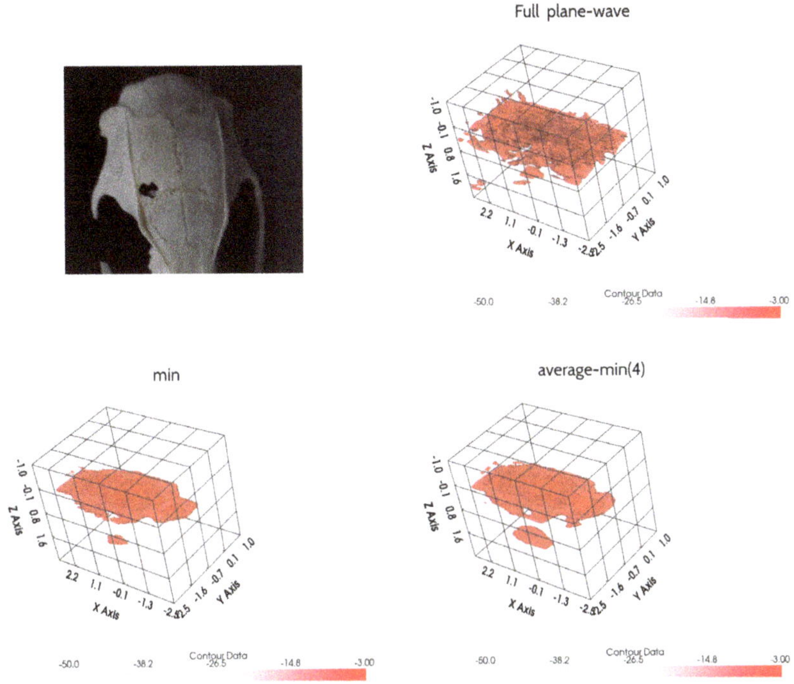

Figure 28. Evaluation of wave penetration through a rat skull using three imaging techniques. A reflector placed behind the skull serves to assess reconstruction performance.

Across all experiments, the proposed *average-min(4)* technique consistently outperforms both the *full plane wave* and *min* approaches in terms of image quality, spatial resolution, and noise suppression. In both single-channel and multi-element signal analysis, it demonstrates improved clarity and reduced interference. When imaging closely spaced reflectors and structures at varying depths, it achieves superior axial and lateral resolution, as confirmed by contour analysis at -6 dB and -12 dB. The method also proves effective in challenging scenarios, such as imaging through bone tissue, where it maintains spatial fidelity and accurately reconstructs reflectors behind complex media. These results validate the robustness and versatility of planar wave imaging based on spatial sparse diversity for high-resolution ultrasonic applications. Additionally, the reception process can benefit from beamforming-based processing techniques that enhance image contrast by suppressing secondary lobes [26–30].

7. Discussion and Conclusions

This study analyzes the performance of a two-dimensional array with large elements for very near-field imaging, specifically targeting applications in preclinical cerebrovascular research. Under these conditions, our interest lies in ensuring that the aperture operates effectively and uniformly, supports high-speed acquisition modes, and remains adaptable to systems with a limited number of active channels.

Metrics were obtained to evaluate the contribution of individual elements to image formation, and optimal operating conditions were proposed to maximize resource efficiency. Based on the results, and considering certain trade-offs, a viable working range is defined from 20 mm and up to 30 mm, with no more than 20 mm of coupling material.

Using region-specific sparsity strategies, a novel methodology was developed to exploit transmit aperture diversity and significantly suppress sidelobes generated by plane wave transmission. Unlike conventional approaches that rely on a single optimized solution, the proposed method employs multiple transmit apertures randomly selected within the constraints of each strategy, while maintaining a common receive aperture.

Simulations demonstrate that applying minimization criteria to samples corresponding to the same receive elements but different transmissions effectively reduces noise, suppresses secondary oscillation patterns, and enhances echo bandwidth, thereby improving axial resolution. This approach eliminates the need for transmit beam steering and enables full operation in plane wave mode without introducing transmit delays. Additionally, it offers substantial design flexibility for the receive aperture, which is no longer constrained by grating lobe generation.

Image quality, particularly in terms of lateral resolution and contrast enhancement, depends on the design of the receive aperture and the beamforming techniques employed. A technique combining two complementary apodizations has been proposed, yielding results comparable to those obtained with a full aperture. The core concept of this work has been experimentally validated using a smaller aperture with a similar design across various scenarios, including volumetric imaging of a rat skull.

The experimental figures reviewed confirm the practical feasibility of the proposed methodology, demonstrating consistent improvements in image contrast across different conditions and validating the theoretical predictions regarding axial resolution and overall image quality.

The analysis indicates that dividing the image area into two regions and adapting the transmission accordingly yields improved performance. However, performance varies between regions due to modulation effects introduced by the elements. The central region achieves the best results, exhibiting a contrast 6dB higher than that of the outer region. These differences can be compensated during acquisition so that both regions are combined in the pre-beamforming stage, providing a more natural image and reducing data volume. However, this also transfers secondary lobe patterns between regions and may reduce the overall dynamic range. Independent processing of both regions minimizes mutual interference, but a dedicated processing technique is required to effectively merge them.

The technological implementation of this solution involves addressing several key challenges. First, depending on the selected operational mode, the system requires the development of two to three independent multiplexing structures: two for transmission (potentially shared with reception) and one specifically for reception. The first structure uses a Taylor apodization scheme; the second employs a ring-distribution scheme; and the third, dedicated to reception, is designed to cover the full aperture in a maximum of four firings.

Second, since beamforming does not require knowledge of the positions of the transmitting elements, the system could incorporate a mechanism to autonomously and randomly select transmit apertures within the constraints of each transmission strategy. Additionally, a processing stage could be implemented in the receiving channels to apply a nonlinear, minimum-based filter functioning as an EMI-like filtering stage [25].

In summary, this work provides solutions to the low-contrast issue in plane wave imaging without requiring wavefront deflection. It optimizes acquisition processes, improves data flow efficiency, and reduces computational cost by enabling full image reconstruction using the same data volume as a single acquisition. Furthermore, the proposed approach simplifies the design of sparse apertures involved in the process, particularly by avoiding the need to optimize specific transmission solutions and, by minimizing grating lobe generation, facilitating the design of the receive aperture.

The development of specialized instrumentation based on this operational approach opens new possibilities for designing portable and low-cost imaging systems, especially in preclinical research environments where flexibility and efficiency are essential.

Author Contributions: Conceptualization and methodology, Ó.M.-G.; software, J.H.; formal analysis, investigation and validation, Ó.M.-G. and L.E.; resources, M.P., G.C. and J.C.; writing, Ó.M.-G.; review and comments, G.C., J.C., Ó.M.-G. and L.E.; experiments, Ó.M.-G., G.C. and J.C. All authors have read and agreed to the published version of the manuscript.

Funding: Jorge Huecas and Guillermo Cosarinsky, Staff hired under the Generation D initiative, promoted by Red.es, an organisation attached to the Ministry for Digital Transformation and the Civil Service, for the attraction and retention of talent through grants and training contracts, financed by the Recovery, Transformation and Resilience Plan through the European Union's Next Generation funds. This research work was funded by the European Commission–NextGenerationEU, through Momentum CSIC Programme: Develop Your Digital Talent. This research work was funded by: PID2022-138013OB-I00, MCIN/AEI /10.13039/501100011033/FEDER, UE. This research work was funded by: PID2022-143271OB-I00, MCIN/AEI /10.13039/501100011033/FEDER, UE. This work has been developed within the framework of the TEC-2024/TEC-43 LUNABRAIN-CM project, funded with 1.026 million by the Community of Madrid through the R&D activities program in technologies, granted by Order 5696/2024.

Institutional Review Board Statement: Not applicable. This study was conducted within a theoretical framework, prior to experimentation, and does not involve the use of animal models.

Data Availability Statement: The datasets presented in this article are not readily available as they are part of an ongoing study. Requests to access the datasets should be directed to the corresponding author. The data will be available to the public without restrictions at https://digital.csic.es accessed on 10 January 2029.

Conflicts of Interest: The authors declare no conflicts of interest. The funders had no role in the design of the study; in the collection, analyses, or interpretation of data; in the writing of the manuscript; or in the decision to publish the results.

References

1. Provost, J.; Papadacci, C.; Demene, C.; Gennisson, J.L.; Tanter, M.; Pernot, M. 3D ultrafast Doppler imaging applied to the noninvasive mapping of blood vessels in vivo. *IEEE Trans. Med. Imaging* **2016**, *35*, 1109–1117.
2. Errico, C.; Pierre, J.; Pezet, S.; Desailly, Y.; Lenkei, Z.; Couture, O.; Tanter, M. Ultrafast ultrasound localization microscopy for deep super-resolution vascular imaging. *Nature* **2015**, *527*, 499–502. [CrossRef]
3. Chavignon, A.; Heiles, B.; Hingot, V.; Orset, C.; Vivien, D.; Couture, O. 3D Transcranial Ultrasound Localization Microscopy in the Rat Brain with a Multiplexed Matrix Probe. *IEEE Trans. Biomed. Eng.* **2022**, *69*, 2132–2142. [CrossRef] [PubMed]
4. Provost, J.; Garofalakis, A.; Sourdon, J.; Bouda, D.; Berthon, B.; Viel, T.; Perez-Liva, M.; Lussey-Lepoutre, C.; Favier, J.; Correia, M.; et al. Simultaneous positron emission tomography and ultrafast ultrasound for hybrid molecular, anatomical and functional imaging. *Nat. Biomed. Eng.* **2018**, *2*, 85–94. [CrossRef] [PubMed]
5. Zhang, C.; Lei, S.; Ma, A.; Wang, B.; Wang, S.; Liu, J.; Shang, D.; Zhang, Q.; Li, Y.; Zheng, H.; et al. Evaluation of tumor microvasculature with 3D ultrasound localization microscopy based on 2D matrix array. *Eur. Radiol.* **2024**, *34*, 5250–5259. [CrossRef]
6. Coudert, A.; Denis, L.; Chavignon, A.; Bodard, S.; Naveau, M.; Sistiaga, P.P. 3-D Transcranial Ultrasound Localization Microscopy Reveals Major Arteries in the Sheep Brain. In *IEEE Transactions on Ultrasonics, Ferroelectrics, and Frequency Control*; no. 12: Breaking the Resolution Barrier in Ultrasound; IEEE: New York, NY, USA, 2024; Volume 71, pp. 1666–1676. [CrossRef]
7. Chabouh, G.; Denis, L.; Bodard, S.; Lager, F.; Renault, G.; Chavignon, A.; Couture, O. Whole Organ Volumetric Sensing Ultrasound Localization Microscopy for Characterization of Kidney Structure. *IEEE Trans. Med. Imaging* **2024**, *43*, 4055–4063. [CrossRef]
8. Li, X.; Gachagan, A.; Murray, P. Design of 2D Sparse Array Transducers for Anomaly Detection in Medical Phantoms. *Sensors* **2020**, *20*, 5370. [CrossRef] [PubMed]
9. Martínez-Graullera, Ó; de Souza, J.C.E.; Parrilla Romero, M.; Higuti, R.T. Design of 2D Planar Sparse Binned Arrays Based on the Coarray Analysis. *Sensors* **2021**, *21*, 8018. [CrossRef]

10. Maffett, R.; Boni, E.; Chee, A.J.Y.; Yiu, B.Y.S.; Savoia, A.S.; Ramalli, A.; Tortoli, P.; Yu, A.C.H. Unfocused Field Analysis of a Density-Tapered Spiral Array for High-Volume-Rate 3-D Ultrasound Imaging. *IEEE Trans. Ultrason. Ferroelectr. Freq. Control* **2022**, *69*, 2810–2822. [CrossRef]
11. Saint-Martin, L.; Avanaki, K. Skull-Mimicking Phantoms with Tunable Properties for Transcranial Ultrasound Imaging. *IEEE Access* **2025**, *13*, 79503–79515. [CrossRef]
12. Montaldo, G.; Tanter, M.; Bercoff, J.; Benech, P.; Fink, M. Coherent plane-wave compounding for very high frame rate ultrasonography and transient elastography. *IEEE Trans. Ultrason. Ferroelectr. Freq. Control* **2009**, *56*, 489–506. [CrossRef]
13. Jacquet, J.B.; Guey, J.L.; Kauffmann, P.; Tamraoui, M.; Roux, E.; Nicolas, B.; Coffy, E.; Liebgott, H. Simulation, Design and characterization of a Large Divergent Element Sparse Array (LDESA) for 3D Ultrasound Imaging. *IEEE Trans. Ultrason. Ferroelectr. Freq. Control* **2025**, *72*, 1041–1052. [CrossRef]
14. Heiles, B.; Correia, M.; Hingot, V.; Pernot, M.; Provost, J.; Tanter, M.; Couture, O. Ultrafast 3D Ultrasound Localization Microscopy Using a 32 × 32 Matrix Array. *IEEE Trans. Med. Imaging* **2019**, *38*, 2005–2015. [CrossRef]
15. Jones, R.M.; DeRuiter, R.M.; Lee, H.R.; Munot, S.; Belgharbi, H.; Santibanez, F. Non-invasive 4D transcranial functional ultrasound and ultrasound localization microscopy for multimodal imaging of neurovascular response. *Sci. Rep.* **2024**, *14*, 30240. [CrossRef] [PubMed]
16. Tanter, M.; Fink, M. Ultrafast imaging in biomedical ultrasound. *IEEE Trans. Ultrason. Ferroelectr. Freq. Control* **2014**, *61*, 102–119. [CrossRef]
17. Montaldo, G.; Urban, A.; Macé, E. Functional Ultrasound Neuroimaging. *Annu. Rev. Neurosci.* **2022**, *45*, 491–513. [CrossRef] [PubMed]
18. Koo, S.; Jang, D.; Yu, J.; Yoon, H. Sub-pitch plane-wave imaging for improved 3-D ultrasound imaging with a large pitch 2-D array. *Biomed. Eng. Lett.* **2025**. [CrossRef]
19. Gómez-de Frutos, M.C.; García-Suárez, I.; Laso-García, F.; Diekhorst, L.; Otero-Ortega, L.; Alonso-López, E. Identification of brain structures and blood vessels by conventional ultrasound in rats. *J. Neurosci. Methods* **2020**, *346*, 108935. [CrossRef] [PubMed]
20. Harput, S.; Christensen-Jeffries, K.; Ramalli, A.; Brown, J.; Zhu, J.; Zhang, G. 3-D Super-Resolution Ultrasound Imaging with a 2-D Sparse Array. *IEEE Trans. Ultrason. Ferroelectr. Freq. Control* **2020**, *67*, 269–277. [CrossRef]
21. Brunner, C.; Grillet, M.; Sans-Dublanc, A.; Farrow, K.; Lambert, T.; Macé, E.; Montaldo, G.; Urban, A. A Platform for Brain-wide Volumetric Functional Ultrasound Imaging and Analysis of Circuit Dynamics in Awake Mice. *Neuron* **2020**, *108*, 861–875. [CrossRef]
22. Piwakowski, B.; Sbai, K. A new approach to calculate the field radiated from arbitrarily structured transducer arrays. *IEEE Trans. Ultrason. Ferroelectr. Freq. Control* **1999**, *46*, 422–440. [CrossRef]
23. Lerman, R.I.; Yitzhaki, S. A note on the calculation and interpretation of the Gini coefficient. *Econ. Lett.* **1984**, *15*, 363–368. [CrossRef]
24. Ferreira, E.; Garín, A. Una nota sobre el cálculo del índice de Gini. *Estadística Española* **1997**, *39*, 207–218.
25. Fritsch, C.; Sánchez, T.; Anaya, J.J.; Ibañez, A.; Ullate, L.G.; Parrilla, M.; Izquierdo, M.A.G. A pipelined architecture for high speed automated NDE. In Proceedings of the 1995 IEEE Ultrasonics Symposium. Proceedings. An International Symposium, Seattle, WA, USA, 7–10 November 1995; Volume 1, pp. 833–836. [CrossRef]
26. Yang, J.; Li, J.; Chen, X.; Xi, J.; Cai, H.; Wang, Y. Cross Subaperture Averaging Generalized Sidelobe Canceler Beamforming Applied to Medical Ultrasound Imaging. *Appl. Sci.* **2021**, *11*, 8689. [CrossRef]
27. Shen, C.-C.; Hsieh, P.-Y. Two-Dimensional Spatial Coherence for Ultrasonic DMAS Beamforming in Multi-Angle Plane-Wave Imaging. *Appl. Sci.* **2019**, *9*, 3973. [CrossRef]
28. Yagüe-Jiménez, V.; Ibáñez Rodríguez, A.; Parrilla Romero, M.; Martínez-Graullera, O. Rician Beamforming: Despeckle Method via Coarray Projection Stochastic Analysis. *Appl. Sci.* **2020**, *10*, 847. [CrossRef]
29. Nili, V.A.; Kavehvash, Z.; Mostafavi, S.M.; Shabany, M.; Mehrmohammadi, M. Structured Illumination for Super-Resolution Synthetic Aperture Ultrasound Imaging. *IEEE Sens. J.* 2025, *early access*. [CrossRef]
30. Guan, H.; Jiang, C.; Xu, K.; Ta, D. 3D f-k migration beamformer for full matrix ultrasound imaging using row-column-addressed array. *Ultrasonics* **2026**, *158*, 107809. [CrossRef] [PubMed]

applied sciences

MDPI

Article

Tissue Characterization by Ultrasound: Linking Envelope Statistics with Spectral Analysis for Simultaneous Attenuation Coefficient and Scatterer Clustering Quantification

Luis Elvira *, Carla de León, Carmen Durán, Alberto Ibáñez, Montserrat Parrilla and Óscar Martínez-Graullera

Instituto de Tecnologías Físicas y de la Información, Serrano 144, 28006 Madrid, Spain; cdeleonmoran@hotmail.es (C.d.L.); carmen.duran@csic.es (C.D.); alberto.ibanez@csic.es (A.I.); m.parrilla@csic.es (M.P.); oscar.martinez.graullera@csic.es or oscar.martinez@csic.es (Ó.M.-G.)

* Correspondence: luis.elvira@csic.es; Tel.: +34-915-618-806

Featured Application: Multimodal ultrasound imaging using quantitative ultrasound algorithms.

Abstract: This paper proposes the use of quantitative methods for the characterization of tissues by linking, into a single approach, ideas coming from the spectral analysis methods commonly used to determine the attenuation coefficient with the envelope statistics formulation. Initially, the Homodyned K-distribution model used to fit data obtained from ultrasound signal envelopes was reviewed, and the necessary equations to further derive the attenuation coefficient from this model were developed. To test and discuss the performance of these methods, experimental work was conducted in phantoms. To this end, a series of tissue-mimicking materials composed of poly-vinyl alcohol (PVA) loaded with different particles (aluminium, alumina, cellulose) at varying concentrations were manufactured. A single-channel scanning system was employed to analyse these samples. It was verified that quantitative images obtained from the attenuation coefficient and from the scatterer clustering μ parameter (associated with scatterer concentration) effectively discriminate materials exhibiting similar echo envelope patterns, enhancing the information obtained in comparison with the conventional analysis based on B-scans. Additionally, the implementation of quantitative bi-parametric imaging mappings based on both the μ parameter and the attenuation coefficient, as a means to rapidly visualize results and identify areas characterized by specific acoustic features, was also proposed.

Keywords: attenuation coefficient; envelope statistics; quantitative ultrasound; tissue characterization

1. Introduction

Changes in the acoustic scattering properties of tissues has been correlated with the modification of the physical properties of tissues by disease, as demonstrated in [1] for liver and ocular tissues. Such modifications offer a valuable opportunity to enhance the diagnostic capabilities of ultrasound techniques. In general, these properties exhibit frequency dependence, and their spectral features can provide quantitative information for tissue characterization [2]. However, conventional echo envelope algorithms employed for obtaining B-scan images overlook spectral variations. Consequently, a significant portion of frequency-related information is lost.

To avoid this, substantial efforts have been made to develop methods capable of extracting quantitative information from ultrasound data. These advancements aim to improve conventional B-scan images with additional layers of information concerning the scanned tissues. Several algorithms have attempted to characterize tissue sound attenuation by analysing the spectral content of the waves reflected from different depths [3–5]. Moreover, even from commonly used signal envelopes, a deeper statistical analysis of amplitude distribution can provide quantitative information related to tissue state [6]. This statistical analysis relies on the use of various distribution functions such as Rice, Nakagami, or the Homodyned K-distribution, among others [7]. These functions can be fitted to the amplitude distribution of the echoes scattered from tissues. All these approaches provide quantitative information about tissues, which, when combined with the typical B-scan image, can be correlated with their healthy state.

With advancements in hardware processing power, some of these quantitative techniques have been successfully applied under experimental conditions (in vitro and ex vivo). In a dermatological context, these methods have been employed to measure the acoustic properties of human skin [8] and characterize bacterial infections in skin ulcers [9]. The liver has also been a focus of several studies, such as those on fibrosis [10] and fat content [2]. Furthermore, these quantitative methods have been utilized to characterize tumours in rats, both ex vivo and in vivo [11], as well as in cancerous human lymph nodes ex vivo [12], to evaluate the response of breast cancer to chemotherapy [13], and to study prostate cancer [14]. These techniques have more recently been employed in brain tissues, including the induction of brain apoptosis in the developing mammalian brain [15]. Consequently, major ultrasound imaging manufacturers have started incorporating quantitative tissue characterization packages into their devices [16].

The advent of artificial intelligence methods in ultrasound image analysis [17–19] can also benefit from quantitative ultrasound techniques. Deep learning-based solutions for tissue characterization have already demonstrated their potential in identifying different tissue structures [17,20,21]. Obtaining prior quantitative information from ultrasound data can enhance the utilization of machine learning and deep learning strategies by enriching the numerical information provided to artificial intelligence algorithms. This pre-processing step is expected to facilitate the automatic detection of pathological tissue features, as AI algorithms could be trained with more information. Consequently, new generations of ultrasound imaging systems are anticipated to provide more than just B-scan images of scanned tissues and Doppler analysis of blood flow.

This paper proposes further advancements in quantitative approaches by merging spectral analysis techniques used for obtaining tissue attenuation characteristics with methodological procedures based on envelope statistics. First, a wideband radio frequency (RF) signal scattered by tissues is split into different frequency components using bandpass filtering. A similar strategy has previously been employed by Bigellow [4] for calculating tissue attenuation based on the spectral content of these components. In our study, we employed this signal split into frequency components to conduct statistical analyses at various frequencies after obtaining the signal envelope. It has been shown that one of the parameters obtained by fitting the envelope of scattering signals to the Homodyned K-distribution (commonly referred to as the μ parameter) provides information about scatterer clustering [22,23]. This clustering can be related to scatterer concentration, although the relation is indirect since the μ parameter depends on acquisition system settings, total attenuation, and the density and spatial organization of scatterers [24]. Additionally, it was demonstrated in this paper that the spatial derivative of another parameter from the same distribution (commonly referred to as the σ parameter) can be correlated with the attenuation coefficient. Consequently, all this information can be visualized as new images superimposed onto ultrasound B-scan images, thereby enriching the understanding of tissue characteristics.

The obtained attenuation coefficient can be complemented with information regarding scatterer clustering using the proposed method. The outcome is a 3-mode image comprising the echo envelope, attenuation, and scatterer clustering. Within this work, the algorithm was applied to ultrasound signals from tissue-mimicking phantoms to evaluate the performance of the method.

This new approach may significantly impact clinical settings where the characterization of backscattering spectral changes [11–15] or the Homodyned K-distribution parameters [9,10] has already proven clinically relevant. These include, among others, dermatological, oncological, and hepatic diseases, in which measuring an additional parameter could further enhance the technique's sensitivity. Furthermore, the proposed analysis holds potential for application to other pathologies that would benefit from more precise tissue characterization.

The structure of this paper is as follows: first, the mathematical approach utilizing the Homodyned K-distribution used to analyse ultrasound signals backscattered by tissues was presented in Section 2.1. Next, in Section 2.2, it was shown how the attenuation coefficient can be obtained from the derivative of one of the coefficients provided by this distribution. In the following Section 2.3, the experimental procedure employed to obtain the ultrasound dataset was explained; this section includes the description of the tissue-mimicking samples, the ultrasound hardware used (based on a scanning single-channel transducer arrangement), and the signal treatment methodology. The subsequent Section 3 shows the results obtained and discusses the performance of the presented algorithm in comparison with conventional B-scan images.

2. Materials and Methods

2.1. Envelope Statistics Using the Homodyned K-Distribution

From the point of view of ultrasound wave propagation, biological tissues can be modelled as substrates with randomly or non-randomly distributed sound scatterers. The echo signals produced by this type of media are the superposition of the reflections that take place in each of the scatterers. It has been established that the statistical distribution of the amplitudes of the echo signal envelope carries crucial information about the material responsible for such backscattering [25]. Consequently, this statistical analysis offers a means to characterize tissues and has the potential to provide insights into their health state.

To achieve this characterization, this paper employs the Homodyned K-distribution to fit the amplitude distribution of the echoes backscattered from the tissues. The Homodyned K-distribution is considered one of the most comprehensive scattering models, as it can account for the effect of tissues in scenarios with a limited number of scatterers and can also accommodate periodic amplitude changes within the analysed media [26]. The distribution is expressed by the following integral:

$$p_A(A) = A \int_{u=0}^{\infty} u \frac{J_0(us) J_0(uA)}{\left(1 + \frac{u^2 \sigma^2}{2\mu}\right)^{\mu}} \, du \tag{1}$$

where $p_A(A)$ is the probability density function associated with a particular amplitude A on the signal envelope and $J_0(x)$ is the Bessel function of order 0. This Homodyned K-distribution is described by a set of three parameters: σ, μ, and s. These parameters are associated with the diffuse energy, scatterer clustering, and coherent energy, respectively. It is proposed in this study that variations in the parameter σ, denoted as $\Delta\sigma$, can be linked to changes in the scattered energy ΔE as the ultrasound wave penetrates through the tissues.

From this ΔE, information about the attenuation coefficient of the tissues can be obtained, as elaborated below.

Jakeman [27] and Dutt [28] already demonstrated that the Homodyned K-distribution can be expressed as a discrete sum:

$$E[A^{2r}] = \left(\frac{2\sigma^2}{\mu}\right)^r r! \frac{\Gamma(r+1)}{\Gamma(\mu)} \sum_{p=0}^{r} \frac{\Gamma(\mu + r - p)}{\Gamma(p+1)p!(r-p)!} \left(\frac{s^2\mu}{2\sigma^2}\right)^p \tag{2}$$

where

$$E\left[A^{2r}\right] = \overline{A^{2r}} \tag{3}$$

Here, Γ is the Gamma function, and $E[A^{2r}]$ represents the averages along the time of the even powers of the amplitude and is commonly referred to as the amplitude of even moments (r being a natural number).

To elaborate further, the three first even moments lead to a set of three equations:

$$\left.\begin{array}{r} E\left[A^2\right] = s^2 + 2\sigma^2 \\ E\left[A^4\right] = 8\left(1+\frac{1}{\mu}\right)\sigma^4 + 8\sigma^2 s^2 + s^4 \\ E\left[A^6\right] = 48\left(1+\frac{3}{\mu}+\frac{2}{\mu^2}\right)\sigma^6 + 72\left(1+\frac{1}{\mu}\right)\sigma^4 s^2 + 18\sigma^2 s^4 + s^6 \end{array}\right\} \tag{4}$$

If this nonlinear system of equations can be solved, an estimation of the three parameters that define the Homodyned K-distribution is obtained. These parameters characterize the scattering properties of a given tissue.

The envelope of a signal along time $A(t)$ can be obtained from the original signal by the addition of the in-phase (x_p) and quadrature (x_q) signals:

$$A(t) = x_p(t) + x_q(t) = x_p(t) + jx_p(t) \tag{5}$$

The average along time ($\overline{\ }$) of the first even moment of the envelope could be obtained from the in-phase signal as:

$$E\left[A^2\right] = \overline{x_p(t)^2|1+j|^2} = 2\overline{x_p(t)^2} \tag{6}$$

2.2. Attenuation Coefficient

In this work, we propose a method to obtain the attenuation coefficient (α) of a given material using the Homodyned K-distribution parameters s and σ. From the first equation of system (4), these parameters are directly related to the averaged intensity of the echo envelope (A^2), which is proportional, in turns, to the intensity of the incident wave. The acoustic attenuation causes the wave intensity to decrease exponentially as the wave penetrates the material.

Following the mathematical formulation proposed in [29], the discrete Fourier transform ($Xi(f)$) of the wave ($xi(t)$), backscattered by a given tissue with the transducer at position i, may be expressed as:

$$X_i(f) \propto f^2|H(f)|^2|V_{inc}(f)|e^{-2\alpha_{tot}z_i}F_i(f)D_i(f,\alpha_{loc}) \tag{7}$$

where f is frequency; $H(f)$ is the dimensionless filtering characteristics of the ultrasound source; $|V_{inc}(f)|$ is the amplitude spectrum of the voltage pulse applied to the ultrasound source; z_i is the distance from the ultrasound source to the beginning of the region of interest in the tissue when the transducer is at discrete position i; α_{tot} and α_{loc_i} are the attenuation coefficients of the tissue until reaching the region of interest and within this region of interest, respectively; $F_i(f)$ accounts for the frequency dependent tissue reflectivity; and $D_i(f, \alpha_{loc_i})$ takes into account the diffraction pattern within this region.

If the transducer approaches the tissue a distance Δz, the relation of the spectrum in the new position $i + 1$ and the previous position would be:

$$\frac{X_{i+1}(f)}{X_i(f)} = \frac{e^{-2\alpha_{tot} z_i} e^{-4\alpha_{loci}\Delta z} F_{i+1}(f) D_{i+1}(f, \alpha_{loci})}{e^{-2\alpha_{tot} z_i} F_i(f) D_i(f, \alpha_{loci})} = e^{-2\alpha_{loc_i}\Delta z} \quad (8)$$

where it was assumed that the transducer displacement is small and, consequently, no significant changes in the reflectivity or local attenuation of the tissue are produced, and then $F_{i+1}(f) = F_i(f)$ and $D_{i+1}(f, \alpha_{loci}) = D_i(f, \alpha_{loci})$.

Parceval's theorem relates the time average of a signal squared with the frequency average $\langle\ \rangle$ of its discrete Fourier transform as:

$$\overline{x_p(t)^2} = N\left\langle X(f)^2 \right\rangle \quad (9)$$

where N is the number of samples of the signal in the frequency domain. From (6), (8), and (9), the relationship between the envelope energy at consecutive depths would be:

$$\frac{E_{i+1}[A^2]}{E_i[A^2]} = \frac{2N\left\langle X_{i+i}(f)^2 \right\rangle}{2N\left\langle X_i(f)^2 \right\rangle} = \frac{\left\langle X_i(f)^2 e^{-4\alpha_{loci}\Delta z} \right\rangle}{\left\langle X_i(f)^2 \right\rangle} \quad (10)$$

To solve these frequency averages, some knowledge about the attenuation coefficient is needed. For a narrow bandwidth, the frequency dependence of the local attenuation coefficient in tissues may be approximated by a linear function [4]:

$$\alpha_{loci} = \alpha_{0i} f + c_i \quad (11)$$

where α_{0i} and c_i are the slope and y-intercept, respectively. In addition, approximating the spectrum of the ultrasound beam $X_i(f)$ by a Gaussian function of central frequency f_{c_i} and standard deviation d_i, the frequency average can be integrated.

$$\frac{E_{i+1}[A^2]}{E_i[A^2]} = \frac{\int_{-\infty}^{\infty} A_i^2 e^{-2\left(\frac{f-f_{ci}}{d_i}\right)^2} e^{-4\alpha_{loci}\Delta z} df}{\int_{-\infty}^{\infty} A_i^2 e^{-2\left(\frac{f-f_{ci}}{d_i}\right)^2} df} = \frac{A_i^2 d_i \sqrt{\frac{\pi}{2}} e^{-4\Delta z(\alpha_{0i} f_{c_i} + c_i)} e^{-(2\Delta z \alpha_{0i} d_i)^2}}{A_i^2 d_i \sqrt{\frac{\pi}{2}}} \quad (12)$$

Using a first order approach for moderate attenuation, the last exponential in the numerator, depending on the square of the attenuation, could be approximated to 1, which makes:

$$\frac{E_{i+1}[A^2]}{E_i[A^2]} = e^{-4\Delta z(\alpha_{0i} f_{c_i} + c_i)} \quad (13)$$

Combining this equation with the first system of Equation (4), corresponds to the Homodyned K-distribution model:

$$\frac{s_{i+1}^2 + 2\sigma_{i+1}^2}{s_i^2 + 2\sigma_i^2} = e^{-4\Delta z(\alpha_{0i} f_{c_i} + c_i)} = e^{-4\Delta z \alpha_{loci}(f_{c_i})} \quad (14)$$

And, finally, the local attenuation coefficient at depth i can be extracted from (13) as:

$$\alpha_{loci}(f_{c_i}) = -\frac{1}{4\Delta z} ln \frac{s_{i+1}^2 + 2\sigma_{i+1}^2}{s_i^2 + 2\sigma_i^2} \quad (15)$$

This result shows the way to relate the attenuation coefficient to the parameters defining the Homodyned K-distribution scattering model. Therefore, by analysing the changes in the Homodyned K-distribution parameters (s and σ) along the ultrasound signal's propagation path, the attenuation coefficient as a function of z for a given material can be derived.

In the present paper, ultrasound images used for material characterization were obtained by successive linear scans at different depths performed with a single-channel focused transducer. To obtain the attenuation coefficient in this case, a small correction should be introduced as two different regions (water coupling layer and tissue-mimicking material) are traversed by the ultrasound beam (Figure 1). If the speeds of sound in the coupling medium and in the analysed tissue are similar, wave refraction can be neglected. As a result, when the transducer approaches the tissue, the decrease in the distance travelled by the wave in the coupling material can be assumed to be equal to the increase in the distance travelled through the tissue. This approximation is generally acceptable when using water or coupling gels as coupling media for biological tissues. In that case, when the transducer performs two consecutive linear B-scans (i and $i+1$) with Δz being the variation in distance to the tissue between them, Equation (8) becomes:

$$\frac{X_{i+1}(f)}{X_i(f)} = e^{-2\alpha_{loci}\Delta z} e^{-2(\alpha_{loci}-\alpha_{coupling})\Delta z} \quad (16)$$

where $\alpha_{coupling}$ is the attenuation coefficient in the coupling material. Following a similar strategy as before, the envelopes of these signals are obtained, and replacing their mean square values with the Homodyned K-distribution parameters becomes:

$$\alpha_{loci}(f_{c_i}) = \alpha_{coupling}(f_{c_i}) - \frac{1}{4\Delta z} ln \frac{s_{i+1}^2 + 2\sigma_{i+1}^2}{s_i^2 + 2\sigma_i^2} \quad (17)$$

As noted, this solution was obtained when narrow pulses around a central frequency f_{c_i} are used; therefore, the attenuation frequency dependence can be considered linear. Alternatively, when using wideband pulses and the attenuation frequency dependence becomes nonlinear, a parallel formulation could be followed using the signals resulting from a narrow bandpass filtering of the scattering signals centred at successive frequencies f_j instead of the original signals. A similar strategy was previously proposed by Bigelow [4], where multiple Gaussian filters were applied to the backscattered echoes to obtain frequency shifts from which the attenuation coefficient was calculated using Fourier Transform analysis. Now, different the Homodyned K-distribution parameters s_i and σ_i can be obtained from the envelopes of the signal i filtered at different central frequencies f_j becoming s_{i,f_j} and σ_{i,f_j}, respectively. Therefore, the attenuation coefficient (17) would be different for each frequency band and can be expressed as:

$$\alpha_{loci}(f_j) = \alpha_{coupling}(f_j) - \frac{1}{4\Delta z} ln \frac{s_{i+1,f_j}^2 + 2\sigma_{i+1,f_j}^2}{s_{i,f_j}^2 + 2\sigma_{i,f_j}^2} \quad (18)$$

To provide a more general characterization of tissue attenuation as a function of frequency in the case of wideband pulses, the frequency dependence of the attenuation coefficients obtained at the different central frequencies f_j can be fitted to a power form:

$$\alpha_{loc_i}(f) = \alpha_{0i} f^n + c_i \quad (19)$$

This is a generalization of the common linear formula previously used in Equation (11) for a narrow bandwidth, where n was equal to 1. Fitting this power form to the attenuation coefficient values obtained experimentally at different frequencies improves the reliability and regularizes the results, especially when working with data acquired from ultrasound devices covering a broad frequency spectrum [30].

As a result, the proposed methodology allows us to infer information about the attenuation coefficient of tissues based on variations in the parameters σ and s, which are directly related to changes in scattered intensity.

The subsequent sections of this paper will delve into the experimental methodology and experimental results in order to discuss the performance of the mathematical approach presented.

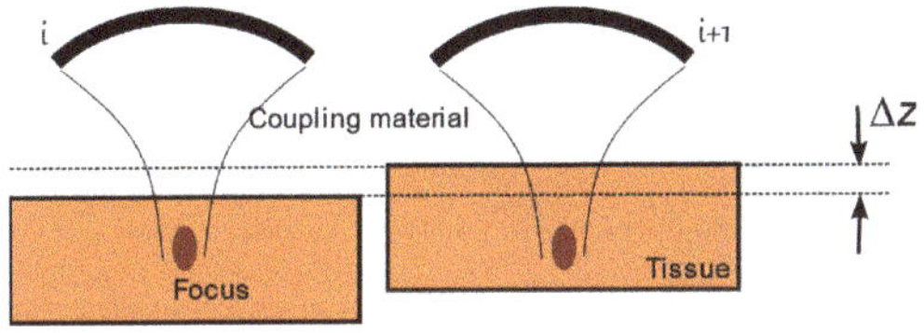

Figure 1. As the distance between the transducer and the material decreases, Δz represents the change in the space crossed by the beam in different materials (coupling material and tissue) between the i and $i + 1$ B-scans.

2.3. Experimental Methodology

Tissue-mimicking phantoms with diverse characteristics were prepared using polyvinyl alcohol (PVA) as described in [31]. That study highlighted the suitability of this gel for phantom fabrication due to its similarity in density and sound speed to biological soft tissues, the robustness of the resulting gels, and the possibility of tailoring their properties through the addition of specific additives to mimic different tissues. The PVA used in this study was 130,000, 99+% hydrolysed (Sigma Aldrich, St. Louis, MI, USA) with a 10% mass concentration in water. Following the aforementioned reference, this mixture resulted in samples with a density of 1040 kg/m^3 and a sound speed of 1541 m/s at ambient temperature (~25 °C). To introduce acoustic scatterers into the material and alter its attenuation coefficient, three types of particles, namely aluminium, alumina (Al_2O_3), and cellulose, were used at different concentrations up to 8% weight per weight (w/w). The particle sizes mainly ranged from 1 to 5 μm for aluminium and alumina, and above 50 μm for cellulose (with the latter exhibiting a pronounced tendency to flocculate). Additionally, a preservative agent, chlorhexidine at 0.05% w/w, was added to improve the durability and stability of the samples.

For this study, parallelepipedal PVA-based samples loaded with particles were produced. Batches of four samples each were made for five different particle concentrations (1%, 2%, 4%, 6%, and 8% w/w) using aluminium and alumina particles. This customization yields isotropic gels with varying scatterer concentrations and attenuation properties. Although biological tissues often exhibit anisotropic behaviour and boundary irregularities, experiments conducted with these phantoms may provide a preliminary assessment of the algorithm's performance prior to further measurements with real tissues. Each specimen's dimensions were set to 22 mm in length, 10 mm in width, and 5 mm in thickness. Figure 2a depicts two of these PVA samples charged with 4% particle concentrations of aluminium (dark grey) and alumina (light grey). The manufacturing of these elements was made following a moulding procedure, with the moulds being made of poly-lactic acid (PLA) using 3D printing. It should be noted that although these samples are inhomogeneous due to their composition of PVA, water, and particles, they will be referred to as single-material samples throughout this work. This is because, at a macroscopic scale, the properties detected by the ultrasound device are not expected to change significantly within each sample.

Additionally, a composite phantom made from a mixture of three different PVA-based materials was manufactured and will be referred to as the composite phantom (Figure 2b). All these materials shared the same PVA 10% base but had varying concentrations of different particles (2% cellulose, 1% aluminium, and 3.5% aluminium), as shown in Figure 2c. The dimensions of the composite phantom were 20 mm × 10 mm × 6 mm.

The ultrasound system used to obtain images of the developed materials is based on a single-channel transducer configuration performing linear scans. The transducer used in the system was manufactured by S-Sharp, and it has a wideband radio frequency ranging from 20 MHz up to 70 MHz. This spherically focalized transducer features a 10 mm curvature radius and a 6 mm diameter active element.

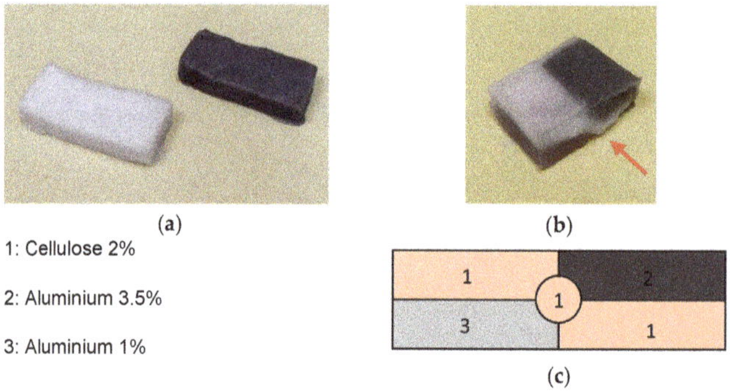

Figure 2. Tissue phantoms based on poly-vinyl alcohol (PVA) made for the experiments. Photograph (**a**) shows parallelepipedic PVA layers loaded with aluminium 4% (dark grey) and alumina 4% (light grey) particles. At the right, in photograph (**b**), the composite phantom made of PVA loaded with aluminium and cellulose particles at different concentrations is shown, and the section view marked by the red arrow is detailed in diagram (**c**).

To ensure accurate positioning of the ultrasound beam focus at the target depth and enable linear scans for image acquisition, the transducer was supported by a motion system consisting of three stages (PI, model Q-521.330) arranged in an XYZ configuration. Each stage is controlled by an independent electronic driver that allows motion in incremental displacements of 0.05 μm along its respective axis. In the present study, a scanning speed of 5 mm/s was used. The ultrasound pulse–echo acquisition system (pulser and digitizer) was synchronized with the horizontal movement to precisely relate signal captures with known transducer positions.

The tissue-mimicking samples were immersed in a water-filled box that acted as the coupling material between the transducer and the samples. The transducer was excited by broadband electric negative pulses emitted by a DPR-500 (JSR Ultrasonics, Pittsford, NY, USA). The echo signals were amplified in the pulser–receiver up to 50 dB and then sent to a Picoscope 6400D digitizer (Pico Technology, Cambridgeshire, UK). A sampling frequency of 313 MHz was used for all acquisitions in this study. Custom software developed in MATLAB 2010b was used to configure all devices, control the movement stages, set signal emission and reception parameters, store the signals acquired by the digitizer, and oversee the entire measurement process.

Given the significant curvature radius of the transducer used, 2D raster scans were performed to change the beam focus position within the samples. After each 10 mm linear scan, the depth was increased in steps of 0.1 mm. A lateral increment of 5 μm between consecutive A-scans allowed for beam superposition to provide image continuity. The acquisition gain was increased as the focus went deeper into the samples to improve the signal-to-noise ratio. Depending on the sample, between 35 and 42 linear scans were required for analysis.

As an example, Figure 3a displays three linear scan images obtained when the focus was placed near the surface, middle, and bottom of the sample, respectively. The upper part of each figure corresponds to the sampled region closest to the transducer. In the first of them, the interface between the water (dark due to the absence of scatterers) and the phantom can be observed. Focussing effects can be seen in all three figures, with the scatterers showing higher amplitude in the middle of the images than above and below the focal plane. The red rectangle in the central image (Scan 23) indicates the area selected to extract the Homodyned K-distribution parameters. Its thickness is 0.75 mm, and it is located near the focal plane. The yellow rectangle shows a 0.1 mm cut, which corresponds to the increases in depth between two successive scans and is used to compose the global B-scan image of the sample (Figure 3b).

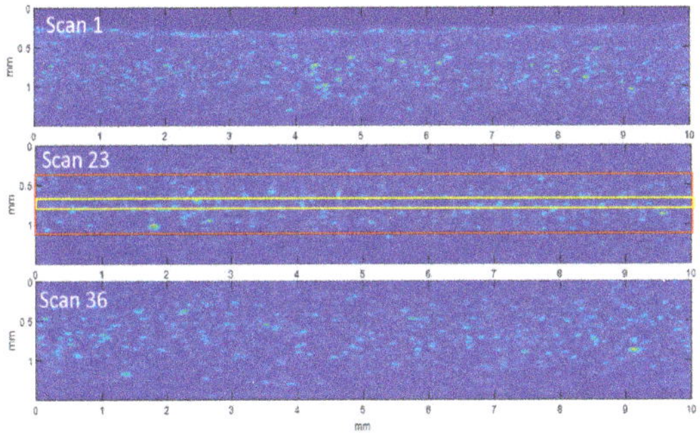

(**a**)

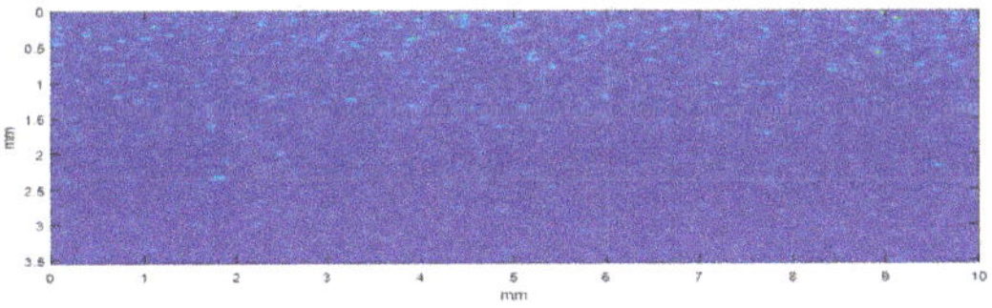

(**b**)

Figure 3. Three different linear B-scans are shown in (**a**) at the upper, middle, and lower parts of the phantom. The red and yellow squares correspond to the signal windows used to calculate the Homodyned K-distribution parameters and the reflectivity global image, respectively. This last reflectivity image is shown in (**b**).

The upper and lower interfaces of the samples were discarded from the image to compound the B-scan of the sample, for further comparison with the images obtained using the Homodyned K-distribution parameters. In this compounded image (Figure 3b), the focussing effect observed in the previous images (Figure 3a) has disappeared. However, the attenuation effect of the wave as it travels through the samples can be appreciated, with the intensity of the scatterers inside the sample decreasing from top to bottom.

To obtain 2D images from the Homodyned K-distribution parameters, the following processing steps were carried out:

(a) Narrow bandpass filtering: Acquisitions were subjected to a battery of 10 MHz bandwidth bandpass filters centred at 20 MHz, 25 MHz, 30 MHz, 35 MHz, and 40 MHz, resulting in five sets of signals representative of these frequency bands.

(b) Calculation of the Homodyned K-distribution parameters for each band: For each frequency band, windows with lateral and axial dimensions of 1 mm and 0.75 mm, respectively, were used to extract the envelopes of the signals along the material. The Homodyned K-distribution parameters were then calculated through a two-step process: Initially, a numerical estimation of the Homodyned K-distribution parameters was obtained by solving the system of Equation (4) using the nonlinear equation solver *lsqnonlin* (MATLAB). This estimation served as the initial value for refining the parameter values on a following step. For this, the same nonlinear equation solver was used, but now envelope data were fitted directly to the probability density function of Equation (1). This two-step process was repeated for all frequencies. Then, a new window was selected with a 0.5 mm lateral displacement for calculating a new set of the Homodyned K-distribution parameters. After completing one row, the procedure continued with the following linear scan until the estimation of the parameters was made for the whole image.

(c) Smoothing of the Homodyned K-distribution parameters: To obtain the attenuation coefficient of the sample, which is based on derivatives with respect to the propagation beam direction, the Homodyned K-distribution parameters were smoothed along the axial direction using splines.

(d) Calculation of the attenuation coefficient for each frequency band: Based on Equation (18), the attenuation coefficient values along the tissue were calculated for each frequency band.

(e) Calculation of the general attenuation coefficient as a function of frequency: Equation (19) was employed to derive the expression of the attenuation coefficient as a function of frequency, utilizing the attenuation coefficients obtained at different frequencies.

The focus of the tissue characterization discussion was put on the analysis of both the attenuation coefficient (emerging from step (e)) and the scatterer clustering parameter (directly related to the μ value which is obtained in step (c)).

3. Results and Discussion

3.1. Analysis of Single-Material Samples

3.1.1. Echo Envelope Images

The analysis of the images obtained from the scanning of the single-material samples was initially performed using the conventional ultrasound imaging method based on the echo envelope. Figure 4 displays sections of 10 different samples of PVA loaded with alumina and aluminium particles, each with varying mass concentrations ranging from 1% to 8%.

Upon examination of these B-scan images, no apparent inhomogeneities in particle distribution could be seen along each sample, indicating a relatively uniform distribution of the particles within the PVA matrix. Also, a gradual increase in the echo density—which leads to echo merging and an increase in echo intensity—was observed, as the particle concentration increased for both types of particles. This increase was particularly significant at lower concentrations; however, as the concentration reached higher values, the effect became less evident.

It is important to consider the differences between the materials based on the nature of the particles. Although the sizes of both kinds of particles were similar, the density of aluminium (2700 kg/m^3) is lower than that of alumina (3950 kg/m^3). Consequently, for a given mass concentration, the number of alumina particles will be smaller than the number of aluminium particles. This difference in the number of particles is noticeable in the images of the right column (alumina), where a smaller number of scatterers is observed, especially at higher concentrations. The reflection coefficient of PVA gel–aluminium and

PVA gel–alumina particles are 0.83 and 0.93, respectively; therefore, a significant impact was not expected by these differences of impedance mismatch in the B-scans.

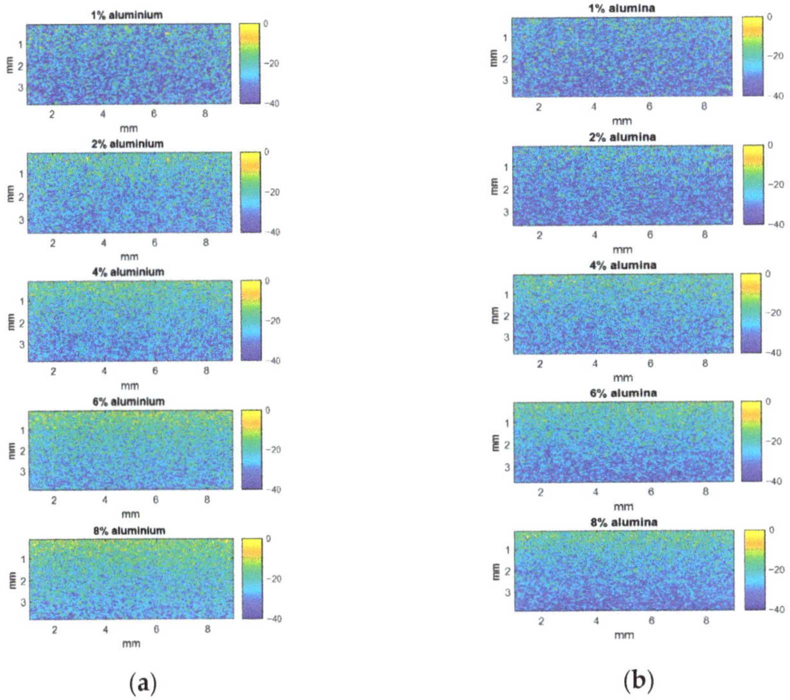

Figure 4. Images of ultrasound reflectivity in decibels corresponding to 10 different samples made with PVA loaded with aluminium (**a**) and alumina (**b**) particles with different concentrations (1%, 2%, 4%, 6%, and 8%) ordered in rows. The decibel scale was referenced to the absolute maximum of the whole set of measurements.

Another important detail observed in these images was the loss of scatter reflections as the ultrasound wave penetrated deeper into the samples, moving from the upper to the lower region. Additionally, the effect became more evident as the particle concentration increased, suggesting a correlation between the concentration of scatterers and the attenuation properties of the materials.

In summary, the echo envelope analysis provided valuable insights into the scattering characteristics of the single-material samples. The images allowed the visualization of particle concentration variations, with higher concentrations leading to more pronounced echo density. Additionally, the observation of the loss of scatter reflections provided information on the attenuation properties of the materials, with aluminium particles showing a more significant effect compared with alumina particles. These results laid the foundation for further quantitative tissue characterization using the Homodyned K-distribution technique.

3.1.2. Attenuation Coefficient Images

Figure 5 presents the attenuation coefficient images obtained from the same 10 samples analysed in the previous section but now utilizing the method proposed in this paper. The images are organized based on frequency (rows), particle concentration (columns), and the nature of the particles used to load the PVA samples: aluminium for Figure 5a and alumina for Figure 5b. The colour mappings reveal a clear increase in the attenuation coefficient with both frequency and particle concentration. The attenuation increment with concentration becomes more evident at the highest frequencies. Notably, the images corresponding to samples loaded with aluminium particles display a higher attenuation coefficient, as evidenced by the faster transition to green and yellow colours (400–600 Np/m).

The echo envelope gradients observed in Figure 4, which were challenging to assess quantitatively in that figure, were transformed into quantitative values of the attenuation coefficient in Figure 5. This transformation allowed for the discrimination between different concentrations, especially in the case of samples loaded with aluminium particles.

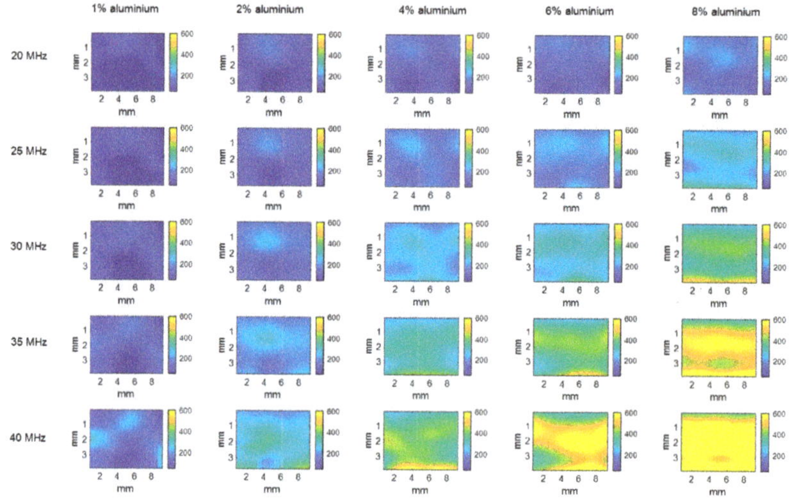

(**a**)

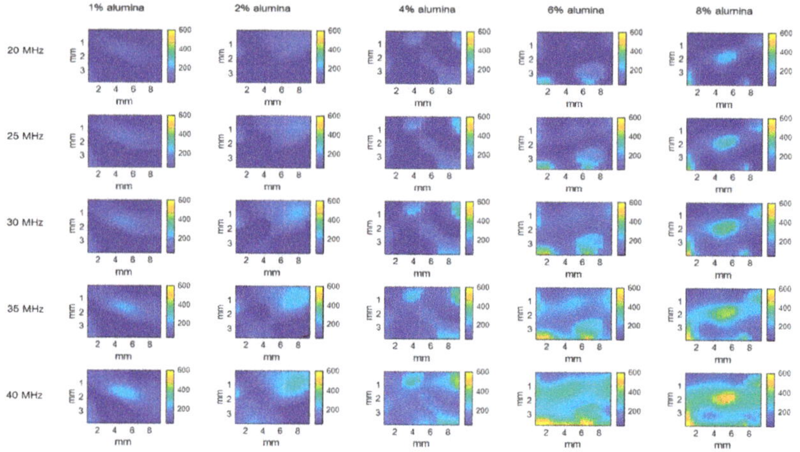

(**b**)

Figure 5. Images of the attenuation coefficient corresponding to 10 different samples made with PVA loaded with aluminium (**a**) and alumina (**b**) particles with different concentrations (1%, 2%, 4%, 6%, and 8%) ordered in columns. The rows show data obtained for different frequencies (20 MHz to 40 MHz in steps of 5 MHz).

Furthermore, a more detailed quantitative analysis of the attenuation coefficient was conducted with data averaged across each image and also averaged for all samples made for each material type (Figure 6). Error bars were plotted to account for the standard deviations of the attenuation coefficient measurements along each sample. Observations made earlier about the increasing attenuation coefficient with increasing frequency and particle concentration are more easily quantified in these plots. Data dispersion is linked to the absolute values of the attenuation coefficient, where similar attenuation coefficients with different particles, frequencies, or concentrations exhibit error bars of the same order. A noteworthy finding is that the attenuation coefficient is better suited for discriminating different materials (particle concentrations) when using higher frequencies, as the overlapping of error bars decreases with an increase in the frequency. This advantage is balanced by the limitation that higher frequencies have lower penetration into tissues. However, for the materials and thicknesses analysed in this work, this effect did not significantly impact the reliability of the obtained attenuation coefficients.

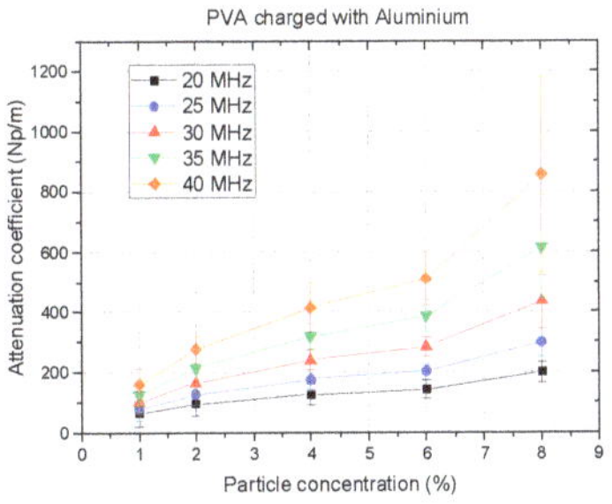

(**a**)

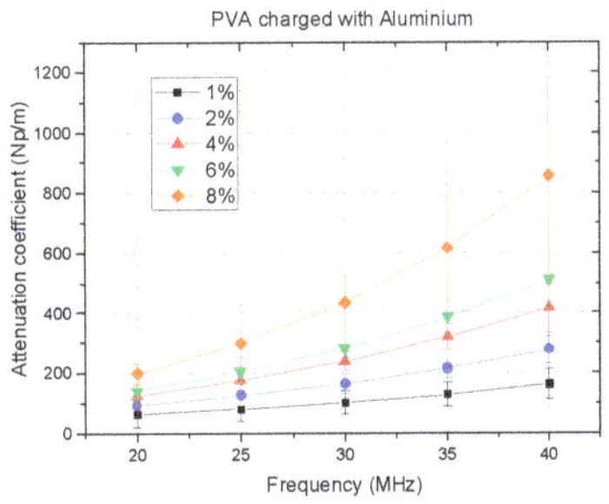

(**b**)

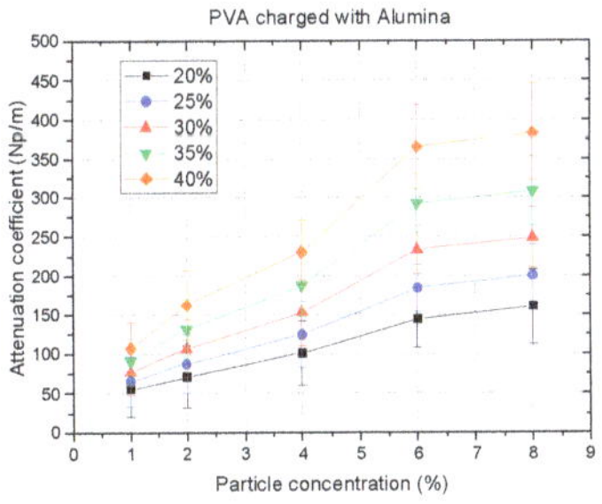

(**c**)

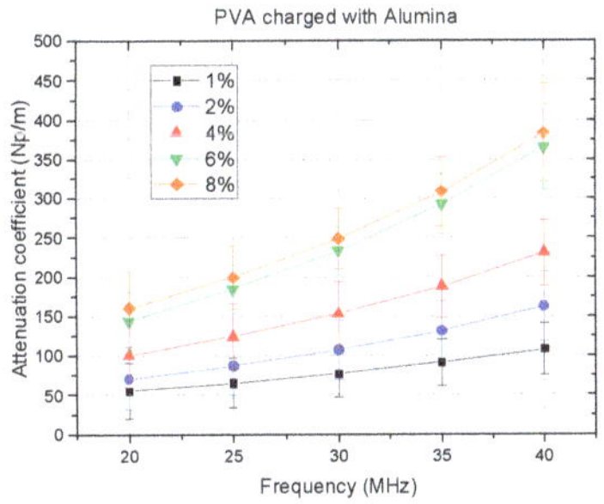

(**d**)

Figure 6. Plots of the attenuation coefficient as a function of particle concentration and frequency averaged along the whole sample. Upper (**a**,**b**) and lower (**c**,**d**) plots correspond to the PVA charged with aluminium and alumina, respectively. At the left (**a**,**c**), the attenuation was obtained as a function of the concentration and, at the right (**b**,**d**), as a function of the frequency.

3.1.3. Scatterer Clustering Parameter Images

The scatterer clustering in the samples was evaluated through the μ parameter obtained from fitting the Homodyned K-distribution to the ultrasound data acquired in each region. Figure 7 displays the images obtained from this parameter for the same 10 samples as before. The trend of the μ parameter to increase with particle concentration is evident for both types of particles. Additionally, materials loaded with alumina particles exhibited lower μ values than those with the same concentrations of aluminium particles, which aligns with the echo envelope results presented earlier and is attributed to the lower density and, consequently, the larger number of aluminium particles.

Frequency also had an impact on the μ parameter, with a moderate decrease observed as the frequency increased. This dependence is related to the different field of view of a wideband transducer as a function of frequency. Considering the resolution cell volume as the volume of the medium where the amplitude of the scattering decreases 3 dB from its maximum, for a 10 mm scan, this volume goes from 0.20 μL at 20 MHz to 0.11 μL at 40 MHz for the transducer used in this work. Then, the resolution cell volume decreased with higher frequency, resulting in fewer echoes detected in the image. However, this effect could not be compensated easily because the μ parameter was also affected by noise, especially at low scatterer concentrations, leading to a "saturation" for the smallest values of μ. This effect was already shown by Fernandez et al. [22], and it is frequency dependent in a complex way.

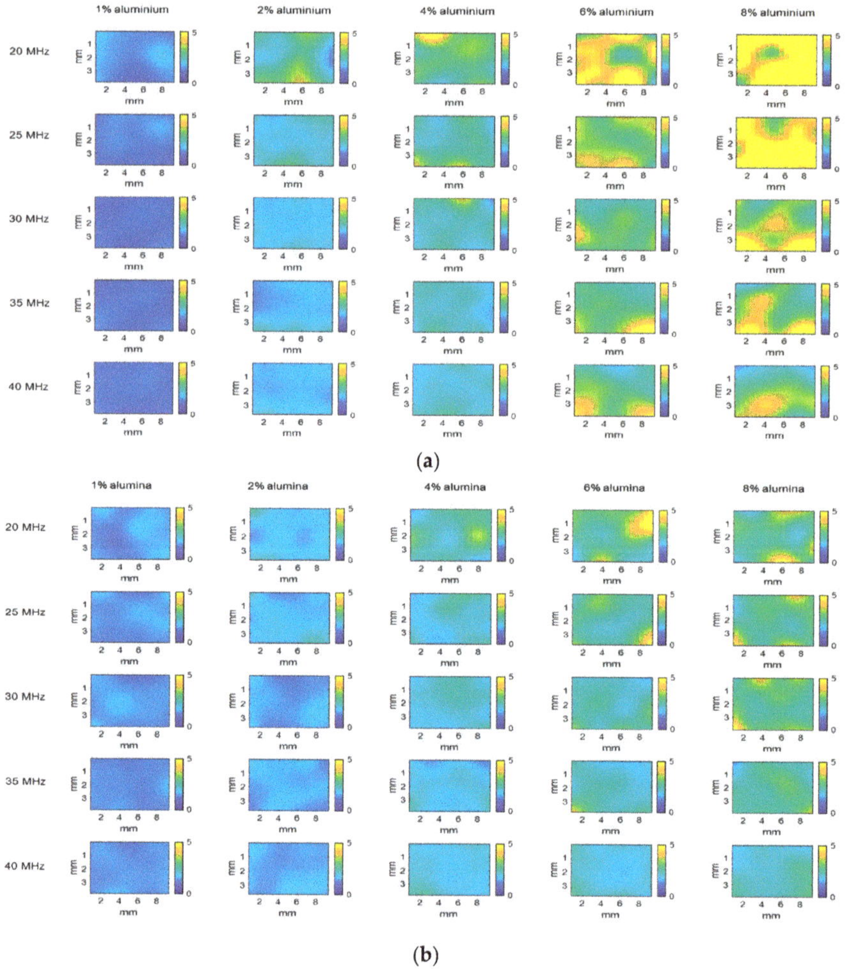

Figure 7. Images of the μ parameter corresponding to 10 different samples made with PVA loaded with aluminium (**a**) and alumina (**b**) particles with different concentrations (1%, 2%, 4%, 6%, and 8%) ordered in columns. The rows show data obtained for different frequencies (20 MHz to 40 MHz in steps of 5 MHz).

Figure 8 demonstrates that the lowest values of μ were obtained for the lowest concentrations and the highest frequencies. Consequently, while the clustering μ parameter is related to scatterer concentration, establishing a simple proportionality between this parameter and the concentration is not straightforward for most cases. Images in Figure 7 also reveal variability in the μ parameter within the samples, similar to what was observed for the attenuation coefficient.

Averaged μ parameter results for all samples with corresponding error bars are shown in Figure 8. The overlapping of error bars was more pronounced for the highest particle concentrations and frequencies. Therefore, contrary to the behaviour of the attenuation coefficient as a descriptor, the μ parameter appears better suited for discriminating materials at low scatterer concentrations and for using the lowest frequencies.

Comparing the behaviour of both the attenuation coefficient and the μ parameter as a function of particle concentration (Figure 6a,c and Figure 8a,c), it can be observed that changes in the slope of the parameters with changes in concentration were consistent for both of them. For example, a change from 6% to 8% concentration in the case of the aluminium-charged samples caused a steep increase in the slope compared with the overall stable trend at lower concentrations, as can be seen in Figures 6a and 8a. This steeper slope appears for both the attenuation coefficient and the μ parameter. The same coherence among these parameters occurs, but with a decrease in the steepness of the slope when there is a change from 6% to 8% concentration in the case of the alumina-charged materials (Figures 6c and 8c); both attenuation coefficients register the same behaviour. This coherence suggests that variations in the expected slope might be the result of local changes in the concentration of particles in the material during the curing process.

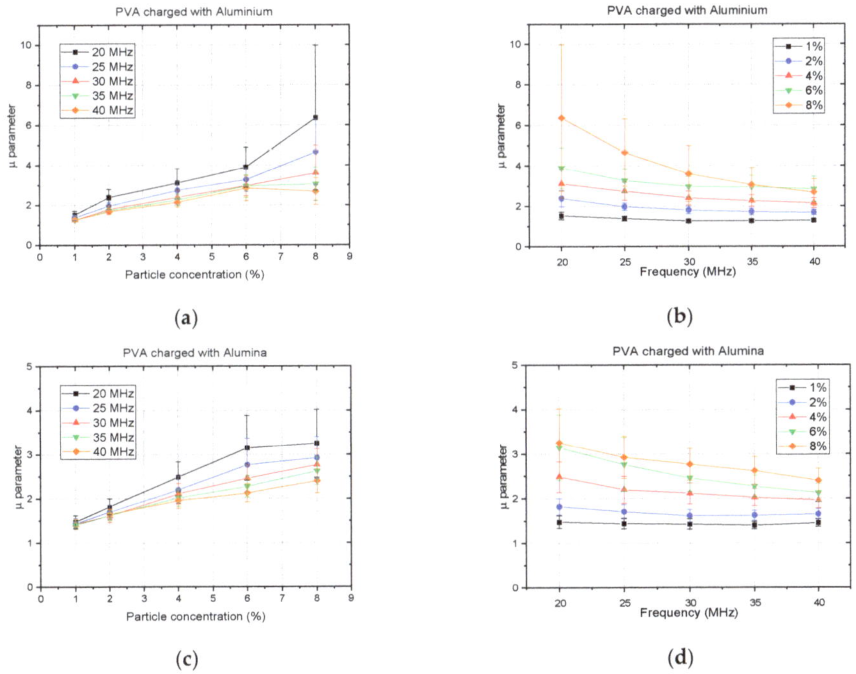

Figure 8. Plots of the μ parameter as a function of particle concentration and frequency averaged along the whole sample. Upper (**a**,**b**) and lower (**c**,**d**) plots correspond to the PVA charged with aluminium and alumina, respectively. At the left (**a**,**c**), the attenuation was obtained as a function of the concentration and, at the right (**b**,**d**), as a function of the frequency.

3.2. Multimodal Image of the Composite Phantom

Unlike the single-material samples, the composite phantom was intentionally designed to include variations in acoustic characteristics, simulating the presence of interfaces between different tissues. This more realistic setup allowed for the study of the performance of the proposed algorithms in scenarios where tissue boundaries were present.

The echo envelope analysis of the composite phantom (Figure 9) was performed using a conventional black and white palette similar to medical B-scans. The boundaries between materials 1 and 3 were barely visible in the centre of the right side of Figure 9a as a horizontal interface, while no noticeable limit was observed between materials 1 and 2. The B-scan image of material 1 displayed a shadow at the bottom central part, which might be misinterpreted as another material. However, this shadow was likely the result of the high attenuation of material 1. To adjust the relative position between acquired data and the material boundaries shown in Figure 9b, the interface between materials 1 and 3 was used as a reference. Overall, the B-scan images of all the PVA-based materials forming the phantom appeared quite similar, making discrimination among them a challenging task based solely on this representation.

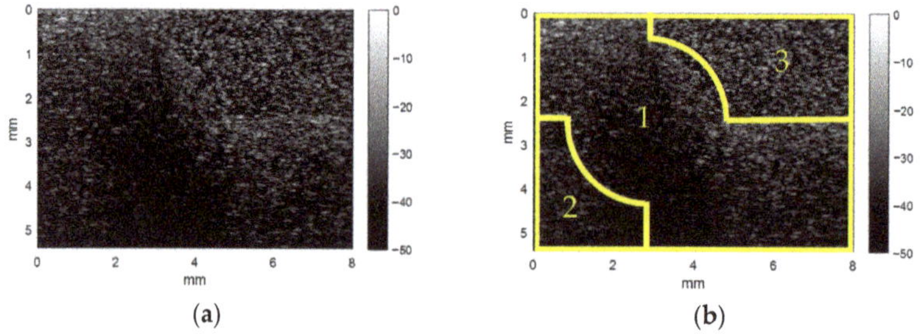

Figure 9. Reflectivity image in decibels without (**a**) and with (**b**) the superimposed diagram of boundaries between the different materials forming the composite phantom, where number 1 corresponds to PVA charged with 2% cellulose particles, number 2 corresponds to PVA charged with 3.5% aluminium particles, and number 3 corresponds to PVA charged with 1% aluminium particles.

The same dataset was then processed using the algorithm to calculate both the attenuation coefficient and the μ parameter (Figure 10). The attenuation coefficient showed a high amplitude region distributed along the diagonal from the top-left to the bottom-right corner, which corresponds to the region occupied by material 1. It also shows a clear low attenuation region at the upper right corner corresponding to material 3. On the other side, the lower left corner (material 2) has a different behaviour as a function of frequency, being closer to the upper right corner at low frequencies and closer to the diagonal region at high frequencies. Based only on attenuation images, it is hard to distinguish materials 2 and 3. The same happens with materials 1 and 3 if only the μ parameter was analysed. This points out the higher efficiency of multiparametric quantitative analysis related to simple models based only on the attenuation coefficient or the μ parameter.

The μ parameter showed a dominant high value in the lower left corner, corresponding to material 2, with a trend of decreasing values towards the upper right corner (material 3), especially at higher frequencies.

The results indicated that the attenuation coefficient was able to clearly distinguish material 1 from materials 2 and 3, while the μ parameter detected a region with a high density of scatterers, corresponding to material 2. Finally, material 3 is clearly different from material 2 in terms of the μ parameter and from material 1 in terms of attenuation. This experiment showcased the capacity of the quantitative algorithm to discriminate among materials. Furthermore, combining both the attenuation coefficient and the μ parameter improved this algorithm's performance, making it more effective in discriminating materials. In contrast, B-scan images are more difficult to interpret and, consequently, more dependent on the observer.

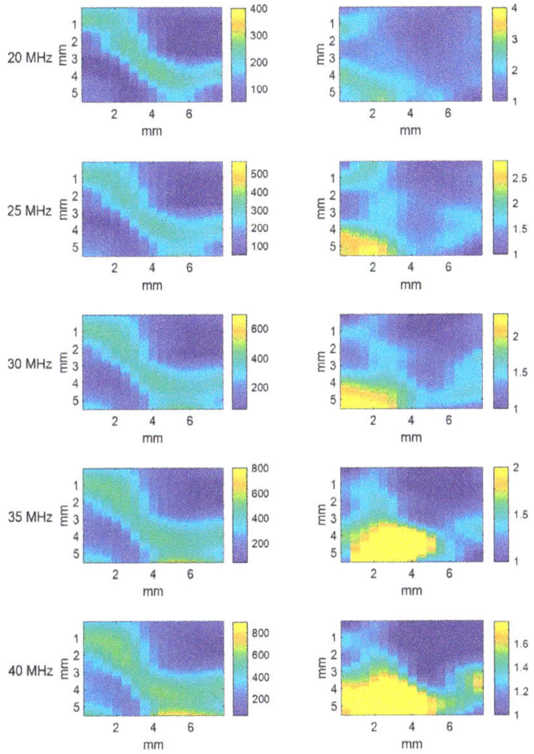

(**a**) (**b**)

Figure 10. Images representing the attenuation coefficient in Np/m (**a**) and the μ parameter (**b**) of the composite phantom. Different rows correspond to the frequencies analysed (20–40 MHz in steps of 5 MHz).

To address the need for an effective representation of both parameters in a single image, a bi-parametric representation was proposed (Figure 11) for the central frequency of 30 MHz. A colour representation is commonly used for other multimodal ultrasound data representation such as Doppler, elastography, etc., which are commonly superimposed to a grey-scaled B-scan image. More recently, when more than two parameters needed to be represented, more complex representations were proposed such as the RGB-based palettes used to represent scattering and reflection characteristics [32,33]. In this work, we chose a palette that could be correlated with the quantitative values of the attenuation coefficient and the μ parameter through a bi-dimensional colour scaling (Figure 11b), which was then plotted with the corresponding tissue images (Figure 11a,c). In this representation, each coloured pixel represented the combination of both parameters, with colours increasing their red content with an increasing attenuation coefficient and their green content with an increasing μ parameter. Colours moving towards black or yellow indicated low or high values of both variables, respectively.

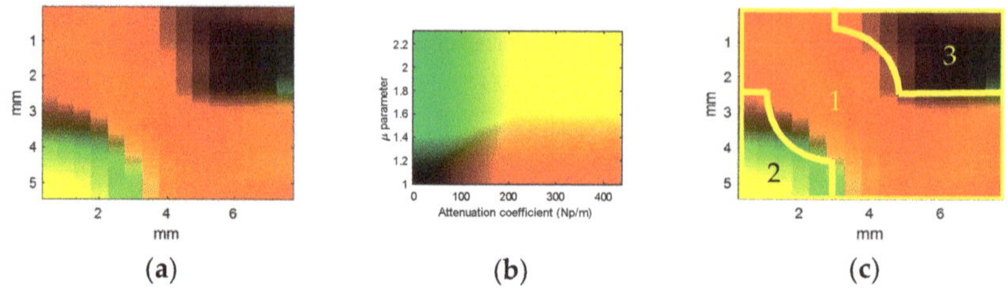

Figure 11. Bi-parametric representation of the attenuation coefficient and the μ parameter of the composite sample for results obtained at 30 MHz (**a**). The corresponding scale, which allows for quantitatively interpreting colours, is shown in (**b**). The same bi-parametric image of (**a**) with the material boundaries superimposed in yellow is presented in (**c**), where number 1 corresponds to PVA charged with 2% cellulose particles, number 2 corresponds to PVA charged with 3.5% aluminium particles, and number 3 corresponds to PVA charged with 1% aluminium particles.

The images resulting from this representation were more complex to analyse quantitatively compared with single-parameter representations. However, they could be used for a quick identification of materials. The dominance of orange in the image corresponded to a μ parameter close to 1.2 and attenuations close to 300 Np/m and clearly identified material 1. On the other side, the yellow region, with a μ parameter above 1.6, and the dark region, with attenuations under 200 Np/m, indicated the locations of the other two materials, 2 and 3, respectively. The bi-parametric representation was found to fit well with the geometrical distribution of the materials, with their boundaries delimited by the yellow-line mask (Figure 11c) being proposed for the first time to represent the μ parameter and the attenuation coefficient together.

Finally, Figure 12 combines this bi-parametric representation with conventional black and white B-scan images, providing transparency to the colour palette belonging to the bi-parametric representation. Although the quantification provided by the colour scaling of Figure 11b was lost, this representation allowed for the application of different information layers that could be shown or hidden by the user within a given tissue image, offering valuable insights into the tissue's characteristics and structure.

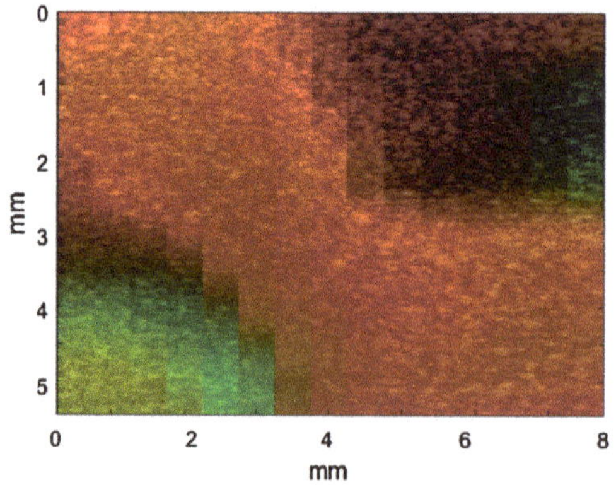

Figure 12. Combined image of the conventional B-scan (in gray scale) with the bi-parametric representation of the attenuation coefficient and the μ parameter obtained from the composite sample for the 30 MHz results. Quantitative parameters are added with transparency to allow the visibility of the B-scan.

4. Conclusions

This paper introduced and demonstrated the use of quantitative methods based on envelope statistics to characterize the acoustic attenuation coefficient and scatterer clustering of tissues together, enhancing the information provided by conventional ultrasound B-scans.

It was shown how envelope statistics based on data fitting to the Homodyned K-distribution can be used to obtain information about the attenuation coefficient. A previous splitting of the wideband radio frequency signal by band filtering at selected frequencies was proposed in order to proceed with further data processing using envelope statistics algorithms.

To assess the performance of the method for the analysis of experimental ultrasound images, an arrangement based on a single-transducer scanning system was used to characterize a set of tissue-mimicking samples manufactured for this study. The transducer incorporated into the system had a 6 mm diameter active element, a spherical focalization with a 10 mm curvature radius, and a wideband radio frequency going from 20 MHz up to 70 MHz. It was excited with a wideband pulser using a pulse–echo working configuration. The transducer was mounted on a precision motion system to perform 2D scans. The samples analysed were made from a PVA–water solution (10% w/w) gels. To make materials with different characteristics, mass concentrations from 1% to 8% of different particles (aluminium, alumina and cellulose) were added to the samples.

B-scan, attenuation coefficient, and μ parameter images were obtained for all the scanned samples. These quantitative parameters were analysed as a function of particle concentration and frequency for the single-material samples, and the standard deviation of the data was also assessed. Results showed that these quantitative parameters could effectively distinguish samples with different particle concentrations, even when subtle variations in echo envelope patterns were present. The attenuation coefficient showed better performance at higher frequencies, while the μ parameter performed better at lower frequencies and concentrations. The study on the composite phantom demonstrated the ability of the proposed methodology to characterize materials with varying characteristics and interfaces. Images of the attenuation coefficient and the μ parameter provided quantitative information about the different materials in the composite phantom. Additionally, a bi-parametric mapping of these magnitudes was proposed, offering a fast identification of the geometrical distributions of different materials in the sample. This representation could be combined with B-scan images to provide tri-modal information about a given tissue.

In conclusion, it was shown that the attenuation coefficient of a scattering material can be obtained from the statistical analysis of envelopes with the help of the Homodyned K-distribution. This approach was used to enhance the information yielded by ultrasound scanning through the generation of multimodal images based on the attenuation coefficient and the clustering of scatterers (by using the single model presented in this work), as well as the conventional tissue echo envelope. The implementation of such quantitative analyses proved successful for the ultrasound characterization of the phantoms developed in this work. However, as noted previously, the inherent anisotropy and boundary irregularities of biological tissues may affect the ability of the algorithms to accurately characterize them. Consequently, further studies on real biological tissues are required to fully assess the potential of the proposed methodology.

Author Contributions: Conceptualization and methodology, L.E.; formal analysis, L.E. and A.I.; investigation, L.E., C.d.L. and C.D.; writing—original draft preparation, L.E.; writing—review and editing, A.I., M.P. and Ó.M.-G. All authors have read and agreed to the published version of the manuscript.

Funding: This work was supported by the projects PID2019-111392RB-I00 and PID2022-138013OB-I00 from the Spanish Ministry of Science and Innovation. This work has been funded with EUR 1.026 million by the Comunidad de Madrid through the LUNABRAIN-CM R&D activities program (TEC-2024/TEC-43), granted by Order 5696/2024.

Data Availability Statement: The original data obtained experimentally and analysed as described in the present study are openly available in the DIGITAL.CSIC repository at http://hdl.handle.net/10261/394298 (accessed on 9 September 2025).

Conflicts of Interest: The authors declare no conflicts of interest.

Abbreviations

The following abbreviations are used in this manuscript:

PVA poly-vinyl alcohol

References

1. Lizzi, F.; Ostromogilsky, M.; Feleppa, E.; Rorke, M.; Yaremko, M. Relationship of ultrasonic spectral parameters to features of tissue microstructure. *IEEE Trans. Ultrason. Ferroelectr. Freq. Control.* **1987**, *34*, 319–329. [CrossRef]
2. Muleki-Seya, P.; Han, A.; Andre, M.P.; Erdman, J.W., Jr.; O'bRien, W.D., Jr. Analysis of two quantitative ultrasound approaches. *Ultrason. Imaging* **2018**, *40*, 84–96. [CrossRef]
3. Labyed, Y.; Bigelow, T.A. Estimating the total ultrasound attenuation along the propagation path by using a reference phantom. *J. Acoust. Soc. Am.* **2010**, *128*, 3232–3238. [CrossRef] [PubMed]
4. Bigelow, T.A. Estimating the total ultrasound attenuation along the propagation path by applying multiple filters to backscattered echoes from a single spherically focused source. *IEEE Trans. Ultrason. Ferroelectr. Freq. Control.* **2010**, *57*, 900–907. [CrossRef]
5. Kim, H.; Varghese, T. Hybrid spectral domain method for attenuation slope estimation. *Ultrasound Med. Biol.* **2008**, *34*, 1808–1819. [CrossRef] [PubMed]
6. Zhou, Z.; Zhang, Q.; Wu, W.; Wu, S.; Tsui, P.-H. Hepatic steatosis assessment using quantitative ultrasound parametric imaging based on backscatter envelope statistics. *Appl. Sci.* **2019**, *9*, 661. [CrossRef]
7. Destrempes, F.; Cloutier, G. A Critical Review and Uniformized Representation of Statistical Distributions Modeling the Ultrasound Echo Envelope. *Ultrasound Med. Biol.* **2010**, *36*, 1037–1051. [CrossRef]
8. Piotrzkowska, H.; Litniewski, J.; Lewandowski, M.; Szymanska, E.; Nowicki, A. Use of quantitative ultrasound to measure acoustic properties of human skin. *Arch. Acoust.* **2009**, *34*, 471–480.
9. Omura, M.; Coron, A.; Bridal, S.L.; Yoshida, K.; Yamaguchi, T. Quantitative ultrasound imaging of bacterial infection on skin ulcers by envelope statistics analysis. In Proceedings of the 2016 IEEE International Ultrasonics Symposium (IUS), Tours, France, 18–21 September 2016; pp. 1–4. [CrossRef]
10. Yamaguchi, T.; Hachiya, H. Proposal of a parametric imaging method for quantitative diagnosis of liver fibrosis. *J. Med. Ultrason.* **2010**, *37*, 155–166. [CrossRef]
11. Bigelow, T.A.; McFarlin, B.L.; O'bRien, W.D.; Oelze, M.L. In vivo ultrasonic attenuation slope estimates for detecting cervical ripening in rats: Preliminary results. *J. Acoust. Soc. Am.* **2008**, *123*, 1794–1800. [CrossRef]
12. Mamou, J.; Coron, A.; Oelze, M.L.; Saegusa-Beecroft, E.; Hata, M.; Lee, P.; Machi, J.; Yanagihara, E.; Laugier, P.; Feleppa, E.J. Three-Dimensional High-Frequency Backscatter and Envelope Quantification of Cancerous Human Lymph Nodes. *Ultrasound Med. Biol.* **2011**, *37*, 345–357. [CrossRef]
13. Sadeghi-Naini, A.; Papanicolau, N.; Falou, O.; Zubovits, J.; Dent, R.; Verma, S.; Trudeau, M.; Boileau, J.F.; Spayne, J.; Iradji, S.; et al. Quantitative ultrasound evaluation of tumor cell death response in locally advanced breast cancer patients receiving chemotherapy. *Clin. Cancer Res.* **2013**, *19*, 2163–2174. [CrossRef]
14. Rohrbach, D.; Wodlinger, B.; Wen, J.; Mamou, J.; Feleppa, E.J. High-Frequency quantitative ultrasound for imaging prostate cancer using a novel micro-ultrasound scanner. *Ultrasound Med. Biol.* **2018**, *44*, 1341–1354. [CrossRef]
15. Rosado-Mendez, I.M.; Noguchi, K.K.; Castañeda-Martinez, L.; Kirvassilis, G.; Wang, S.H.; Manzella, F.; Swiney, B.S.; Masuoka, K.; Capuano, S.; Brunner, K.G.; et al. Quantitative ultrasound and apoptotic death in the neonatal primate brain. *Neurobiol. Dis.* **2019**, *127*, 554–562. [CrossRef] [PubMed]
16. Cloutier, G.; Destrempes, F.; Yu, F.; Tang, A. Quantitative ultrasound imaging of soft biological tissues: A primer for radiologists and medical physicists. *Insights Imaging* **2021**, *12*, 127. [CrossRef]
17. Becker, A.S.; Mueller, M.; Stoffel, E.; Marcon, M.; Ghafoor, S.; Boss, A. Classification of breast cancer in ultrasound imaging using a generic deep learning analysis software: A pilot study. *Br. J. Radiol.* **2018**, *91*, 20170576. [CrossRef]

18. Brattain, L.J.; Telfer, B.A.; Dhyani, M.; Grajo, J.R.; Samir, A.E. Machine learning for medical ultrasound: Status, methods, and future opportunities. *Abdom. Radiol.* **2018**, *43*, 786–799. [CrossRef]
19. Liu, S.; Wang, Y.; Yang, X.; Lei, B.; Liu, L.; Li, S.X.; Ni, D.; Wang, T. Deep Learning in Medical Ultrasound Analysis: A Review. *Engineering* **2019**, *5*, 261–275. [CrossRef]
20. Skandha, S.S.; Nicolaides, A.; Gupta, S.K.; Koppula, V.K.; Saba, L.; Johri, A.M.; Kalra, M.S.; Suri, J.S. A hybrid deep learning paradigm for carotid plaque tissue characterization and its validation in multicenter cohorts using a supercomputer framework. *Comput. Biol. Med.* **2022**, *141*, 105131. [CrossRef] [PubMed]
21. Sahashi, Y.; Kawasaki, M.; Okubo, M.; Kawamura, I.; Kawase, Y.; Yoshida, A.; Tanaka, T.; Hattori, A.; Matsuo, H.; Ozaki, Y. Development of 60 MHz integrated backscatter intravascular ultrasound and tissue characterization of attenuated signal coronary plaques that cause myocardial injury after percutaneous coronary intervention. *Heart Vessel.* **2022**, *37*, 1689–1700. [CrossRef]
22. Fernández, A.; Ibáñez, A.; Parrilla, M.; Elvira, L.; Bassat, Q.; Jiménez, J. Estimation of the concentration of particles in suspension based on envelope statistics of ultrasound backscattering. *Ultrasonics* **2021**, *116*, 106501. [CrossRef]
23. Elvira, L.; Fernández, A.; León, L.; Ibáñez, A.; Parrilla, M.; Martínez, Ó.; Jiménez, J. Evaluation of the Cell Concentration in Suspensions of Human Leukocytes by Ultrasound Imaging: The Influence of Size Dispersion and Cell Type. *Sensors* **2023**, *23*, 977. [CrossRef] [PubMed]
24. Destrempes, F.; Cloutier, G. Statistical modeling of ultrasound signals related to the packing factor of wave scattering phenomena for structural characterization. *J. Acoust. Soc. Am.* **2021**, *150*, 3544–3556. [CrossRef]
25. Oelze, M.L.; Mamou, J. Review of Quantitative Ultrasound: Envelope Statistics and Backscatter Coefficient Imaging and Contributions to Diagnostic Ultrasound. *IEEE Trans. Ultrason. Ferroelectr. Freq. Control.* **2016**, *63*, 336–351. [CrossRef]
26. Destrempes, F.; Cloutier, G. Review of envelope statistics models for quantitative ultrasound imaging and tissue characterization. In *Quantitative Ultrasound in Soft Tissues*; Mamou, J., Oelze, M., Eds.; Springer: Dordrecht, The Netherlands, 2013; pp. 219–274. [CrossRef]
27. Jakeman, E. On the statistics of K-distributed noise. *J. Phys. A Math. Gen.* **1980**, *13*, 31–48. [CrossRef]
28. Dutt, V.; Greenleaf, J.F. Ultrasound echo envelope analysis using a homodyned K distribution signal model. *Ultrason. Imaging* **1994**, *16*, 265–287. [CrossRef]
29. Bigelow, T.A.; Labyed, Y. Attenuation Compensation and Estimation. In *Quantitative Ultrasound in Soft Tissues*; Mamou, J., Oelze, M., Eds.; Springer: Dordrecht, The Netherlands, 2013; pp. 71–93. [CrossRef]
30. Szabo, T.L. Causal theories and data for acoustic attenuation obeying a frequency power law. *J. Acoust. Soc. Am.* **1995**, *97*, 14–24. [CrossRef]
31. Elvira, L.; Durán, C.; Higuti, R.T.; Tiago, M.M.; Ibáñez, A.; Parrilla, M.; Valverde, E.; Jiménez, J.; Bassat, Q. Development and Characterization of Medical Phantoms for Ultrasound Imaging Based on Customizable and Mouldable Polyvinyl Alcohol Cryogel–Based Materials and 3-D Printing: Application to High-Frequency Cranial Ultrasonography in Infants. *Ultrasound Med. Biol.* **2019**, *45*, 2226–2241. [CrossRef]
32. Parker, K.J. Scattering and reflection identification in H-scan images. *Phys. Med. Biol.* **2016**, *61*, L20–L28. [CrossRef] [PubMed]
33. Khairalseed, M.; Brown, K.; Parker, K.J.; Hoyt, K. Real-time H-scan ultrasound imaging using a Verasonics research scanner. *Ultrasonics* **2019**, *94*, 28–36. [CrossRef] [PubMed]

applied sciences

MDPI

Article

How Does Left Ventricular Ejection Fraction Affect the Multimodal Assessment of Congestion in Patients with Acute Heart Failure? Results from a Prospective Study

Laura Karla Esterellas-Sánchez [1,2], Amelia Campos-Sáenz de Santamaría [1,2], Zoila Stany Albines Fiestas [2,3], Silvia Crespo-Aznarez [1,2], Marta Sánchez-Marteles [1,2,4], Vanesa Garcés-Horna [1,2,4], Alejandro Alcaine-Otín [5], Ignacio Gimenez-Lopez [2,4,6,*] and Jorge Rubio-Gracia [1,2,4]

1 Internal Medicine Department, Hospital Clínico Lozano Blesa, 50009 Zaragoza, Spain; lkesterellas@salud.aragon.es (L.K.E.-S.); ameliacampos97@gmail.com (A.C.-S.d.S.); screspo@salud.aragon.es (S.C.-A.); msanchezmar@salud.aragon.es (M.S.-M.); vgarces@salud.aragon.es (V.G.-H.); jrubiogra@posta.unizar.es (J.R.-G.)
2 Aragon Health Research Institute (IIS Aragon), 50009 Zaragoza, Spain; zsalbines@salud.aragon.es
3 Nephrology Department, Hospital Clínico Lozano Blesa, 50009 Zaragoza, Spain
4 School of Medicine, University of Zaragoza, 50009 Zaragoza, Spain
5 Computing for Medical and Biological Applications Group, Faculty of Health Sciences, University San Jorge, 50830 Villanueva de Gállego, Spain; lalcaine@usj.es
6 Aragon Health Sciences Institute (IACS), 50009 Zaragoza, Spain
* Correspondence: igimenez@unizar.es

Abstract: The assessment of systemic congestion in acute heart failure (AHF) remains clinically challenging, particularly across different left ventricular ejection fraction (LVEF) phenotypes. This study aimed to evaluate whether differences exist in the degree of congestion, assessed through a multimodal approach including physical examination, biomarkers (NT-proBNP, CA125), and point-of-care ultrasound using the Venous Excess Ultrasound (VExUS) protocol, between patients with preserved (HFpEF) and reduced ejection fraction (HFrEF). We conducted a prospective observational study involving 90 hospitalized AHF patients, 80 of whom underwent a complete VExUS assessment. Although patients with HFrEF exhibited higher levels of NT-proBNP and CA125, and more frequent signs of third-space fluid accumulation such as pleural effusion and ascites, no statistically significant differences were found in VExUS grades between the two groups. These findings suggest that the VExUS protocol provides consistent and reproducible information on systemic venous congestion, regardless of LVEF phenotype. Its integration into clinical practice may help refine congestion assessment and optimize diuretic therapy. Further multicenter studies with larger populations are warranted to validate its diagnostic and prognostic utility and to determine its potential role in guiding individualized treatment strategies in AHF.

Keywords: heart failure; left ventricular ejection fraction; Venous Excess Ultrasound Score; congestion; biomarkers; NTproBNP; CA125

1. Introduction

Heart failure (HF) is a clinical syndrome characterized by structural or functional cardiac abnormalities that impair ventricular filling or the ejection of blood [1], triggering a cascade of compensatory neurohormonal and inflammatory responses [2–4]. The estimated prevalence of HF is around 2% among adults in developed countries and exceeds 10% in those over 70 years old, making it a major healthcare concern [3]. In Spain, the prevalence is similar [3] and HF remains a leading cause of hospitalization in older adults, primarily due to signs and symptoms of congestion [3,5].

To better understand the clinical course and guide treatment, HF is classified according to left ventricular ejection fraction (LVEF) into three main phenotypes: reduced (HFrEF, ≤40%), mildly reduced (HFmrEF, 41–49%), and preserved (HFpEF, ≥50%) [3]. This classification is clinically significant, as it reflects different underlying etiologies, comorbidities, and, most importantly, variable responses to treatment [5]. HFrEF is characterized predominantly by neurohormonal activation and structural remodeling, whereas HFpEF involves endothelial and microvascular dysfunction and presents a heterogeneous biological profile [5]. Despite these distinctions, decongestive treatment strategies in acute settings often do not differ across these phenotypes [5].

Accurate congestion assessment remains one of the most critical and challenging aspects of HF management. Persistent congestion after hospital discharge is strongly associated with poor prognosis [6]. Hemodynamic congestion, or elevated ventricular filling pressures, often precedes clinical signs by days to weeks [6]. Unfortunately, a reliable non-invasive method for determining volume status is lacking, posing a significant diagnostic challenge when physical examination and standard laboratory tests are inconclusive [7,8].

To address this issue, several tools have emerged, including congestion biomarkers, such as N-terminal pro-B-type natriuretic peptide (NT-proBNP) and Carbohydrate Antigen 125 (CA125) [9,10], and point-of-care ultrasound (POCUS), which has become increasingly integrated into clinical practice across internal medicine units worldwide. Bedside ultrasound has proven to be a valuable diagnostic tool that complements history and physical examination [11–14] and is now considered the fifth pillar of physical examination in HF [15].

Lung ultrasound (LUS) helps detect extravascular pulmonary water through B-lines, which correlate with worse prognosis [13,14]. Inferior vena cava (IVC) diameter assessment provides a simple estimate of right atrial pressure and is associated with changes in pulmonary capillary wedge pressure [12]. Therapy guided by IVC diameter in hospitalized HF patients has been shown to lead to reductions in rehospitalizations and mortality [11–14]. A recent pilot study (CAVAL US-AHF) combining a Doppler assessment of cardiac filling pressures, IVC, and LUS demonstrated that this approach is safe and reliable for guiding therapy [12].

In line with these advances, the recent consensus on the management of fluid overload in acute HF by Pau Llácer et al. [16] proposes two distinct congestion phenotypes and recommends a multiparametric approach that integrates clinical symptoms, biomarkers, and ultrasound findings to guide treatment.

To further refine the quantification of congestion, Beaubien-Souligny et al. proposed the Venous Excess Ultrasound (VExUS) score, a protocol based on Doppler flow patterns in three abdominal veins—hepatic, portal, and intrarenal—that classifies patients into three levels of congestion: none, mild, or severe [17].

The VExUS protocol begins by assessing the IVC. If the diameter is <2 cm, venous congestion is ruled out. If it is ≥2 cm, a further Doppler evaluation is performed. The hepatic vein waveform, portal vein pulsatility, and intrarenal vein pattern are analyzed and integrated into a severity score (grades 0–3). Higher VExUS grades correlate with increased venous pressure and worse outcomes [11,13,18–20].

Although the VExUS score is gaining attention as a tool for assessing systemic venous congestion, it has limitations. These include operator dependency, anatomical variability, and disparities in ultrasound equipment. Moreover, most studies have focused on critically ill patients and those with HFrEF [18,20], leaving its performance across different LVEF phenotypes uncertain.

Some authors suggest that systemic congestion may be less pronounced or may follow a distinct pathophysiological pattern in HFpEF compared to HFrEF [21], which could impact the utility of VExUS in this population. Supporting this, a recent prospective study in HFpEF patients demonstrated that higher VExUS grades correlated with worse clinical outcomes [13,19]. Nevertheless, the pathophysiological characteristics of HFpEF—such as increased ventricular stiffness and diastolic dysfunction—might influence how congestion manifests and, consequently, how accurately VExUS captures it [20].

Although ultrasound is an emerging technique in HF management supported by growing evidence, it still presents certain limitations, including interobserver variability, patient anatomy, and differences in equipment [13,18,22]. Therefore, further studies are needed to explore multimodal congestion assessment strategies and to evaluate the integration of the VExUS protocol in heart failure patients, in order to support or refute its systematic use as a novel tool for patient management. Given these considerations, further research is warranted to determine whether VExUS offers consistent diagnostic value across the full spectrum of LVEF phenotypes.

This study, conducted by the Heart Failure Research Group of the Aragón Health Research Institute (GIIS-043), aims to assess the relationship between VExUS and LVEF and to evaluate the utility of this protocol in real-world clinical settings. Specifically, we aim to determine whether systemic venous congestion quantified by VExUS correlates with clinical, laboratory, and ultrasound findings and whether its diagnostic performance differs according to LVEF.

2. Materials and Methods

2.1. Study Design

This was a prospective, observational, single-center cohort study conducted at the Internal Medicine Department of Hospital Clínico Universitario Lozano Blesa, Zaragoza, Spain. This study included patients from the Cardio-Ren cohort (C.P.—COL21/003) during the 2023–2024 academic year. No pharmacological interventions were performed beyond routine clinical care.

2.2. Study Population

We included adult patients ($\geq$18 years) admitted with a primary diagnosis of acute heart failure (AHF) or decompensated chronic heart failure, who presented with at least one clinical sign or symptom of congestion, such as dyspnea, orthopnea, bendopnea, peripheral edema, pulmonary crackles, ascites, or jugular vein distension, and an elevated NT-proBNP level (>1000 pg/mL) within the first 24 h of admission. This NT-proBNP cutoff was selected based on its frequent use in relevant clinical studies, such as the PARADIGM-HF trial [23], where it identifies patients with higher hemodynamic burden and clinical risk. These inclusion criteria were applied uniformly across all participants to ensure consistency and reduce potential selection bias. Written informed consent was required.

Exclusion criteria included admission from the intensive care unit or emergency observation area with stay > 24 h; refusal to participate or the absence of written consent; severe cognitive or functional impairment precluding outpatient follow-up; and advanced chronic kidney disease (CKD), renal transplant, or current dialysis treatment.

2.3. Data Collection and Variables

The aim of this study was to describe and compare the congestion profiles of patients with preserved versus reduced ejection fraction at the time of admission. Therefore, a multimodal assessment—including physical examination, laboratory tests, and point-of-care ultrasound (POCUS)—was performed within the first 24 h of hospitalization.

We recorded demographic and clinical variables, including age, sex, comorbidities, New York Heart Association (NYHA) functional class, and chronic home medications. Clinical signs of congestion—such as dyspnea, orthopnea, bendopnea, peripheral edema, jugular vein distension, and pulmonary crackles—were systematically assessed. Laboratory tests included assessing the levels of NT-proBNP, CA125, creatinine, sodium, and potassium, and congestion indices such as the combined congestion algorithm (CAC) were calculated.

Echocardiographic evaluation included assessing inferior vena cava (IVC) diameter, the presence of B-lines, pleural or abdominal effusion, and the Venous Excess Ultrasound (VExUS) score, comprising Doppler patterns of hepatic, portal, and intrarenal veins. Congestion was also assessed using a clinical congestion score (range: 0–8) based on dyspnea, edema, and jugular venous distension [7].

Although the primary analysis focused on this baseline evaluation, all patients were monitored throughout hospitalization and prospectively followed for 90 days to assess therapeutic interventions (such as intravenous loop diuretics or hypertonic saline), in-hospital mortality, and post-discharge outcomes including mortality, readmission, and diuretic requirements.

2.4. Echographic Evaluation

All patients underwent standardized bedside ultrasound assessment using Lumify® (Philips Ibérica, Madrid, España) and Omega® (Esaote España S.A.U., Barcelona, Spain) devices, equipped with both phased-array and convex probes. Two trained physicians performed image acquisition after a calibration phase using ten representative cases, achieving interobserver agreement with a kappa coefficient of 0.80 (see Supplementary Materials).

LVEF was obtained from a standard transthoracic echocardiogram performed either during admission or within the previous 12 months. Patients were categorized as having preserved LVEF (HFpEF, ≥50%) or reduced LVEF (HFrEF, <50%).

Lung ultrasound was performed to assess B-lines in eight thoracic quadrants. Pleural effusion and ascites were evaluated via abdominal ultrasound. [IVC diameter was measured approximately 2 cm from its junction with the right atrium, and inspiratory collapsibility was recorded. A collapsibility > 50% was considered normal.].

The VExUS protocol was applied in patients with an IVC diameter ≥ 2 cm. It included a Doppler analysis of the hepatic, portal, and intrarenal veins:

- In the hepatic veins, waveforms were classified as normal, mildly, or severely altered based on the S/D wave ratio and flow direction.
- Portal vein pulsatility was categorized as <30%, 30–50%, or >50%, with higher pulsatility indicating greater congestion.
- Intrarenal venous flow was categorized as continuous, biphasic, or monophasic, with monophasic indicating severe congestion.

Based on these findings, a global VExUS score was assigned, classifying patients into four grades: grade 0 (no congestion), grade 1 (mild), grade 2 (moderate), and grade 3 (severe). For statistical analysis, this classification was dichotomized into low (grades 0 and 1) versus high (grades 2 and 3) systemic venous congestion.

2.5. Statistical Analysis

Qualitative variables were reported as percentages and quantitative variables as means or medians depending on the distribution (assessed by the Shapiro–Wilk test). Group comparisons employed a chi-square test, Student's *t*-test, ANOVA, or the Kruskal–Wallis test, as appropriate. A two-tailed p-value < 0.05 was considered statistically significant. Analyses were performed using Jamovi software (version 2.3.28).

2.6. Ethical Considerations

This study complied with the principles of the Declaration of Helsinki (2013 revision). Approval was granted by the Ethics Committee of the Autonomous Community of Aragón (CEICA). No additional diagnostic procedures were required. Data were pseudonymized using coded identifiers accessible only to the research team. This study was conducted without external funding or participant compensation.

3. Results

A total of 90 patients with (acute heart failure) AHF were included, with a mean age of 84.5 ± 6 years; 48% were women. According to baseline NYHA functional classification, 15.6% were class I, 66.7% class II, 16.6% class III, and 1.1% class IV. Hypertension (86.7%) and atrial fibrillation (75.6%) were the most frequent comorbidities, followed by dyslipidemia (57.8%), CKD (35.6%), diabetes (34.4%), and chronic obstructive pulmonary disease (25.6%). The cohort was predominantly treated with loop diuretics (75.6%), beta-blockers (65.5%), and renin–angiotensin system blockers (53.3%) prior to admission. ISGLT2 use was present in only 28.9% of the sample at the time of the initial assessment. Sixty patients (67%) had HFpEF, and thirty (33%) had HFrEF.

At admission, 35% had orthopnea at rest, 28% with two pillows, and 28% with one. Peripheral edema was moderate-to-severe in 52%, and 83% had jugular vein distention > 6 cm (Figure 1). Clinical congestion scores were similar between groups (median 4 points in HFpEF vs. 5 points in HFrEF; $p = 0.119$).

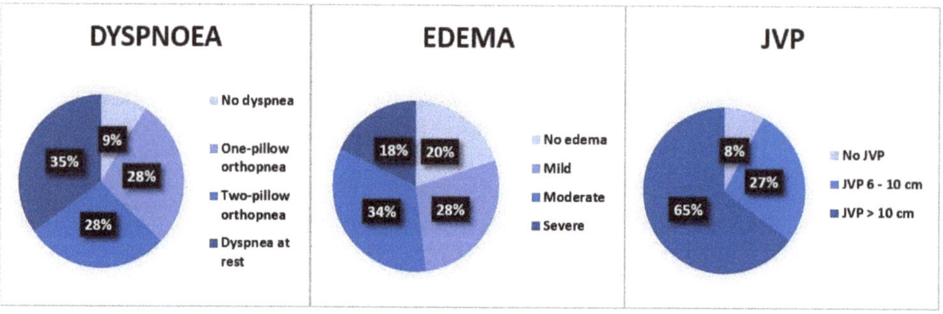

Figure 1. Degree of clinical congestion upon admission.

Lung ultrasound revealed the presence of B-lines in all evaluated patients. These were bilateral in 95% of patients with preserved left ventricular ejection fraction (HFpEF) and in 93% of those with reduced ejection fraction (HFrEF). Pleural effusion was observed in 52% of the HFpEF group, compared to 77% in the HFrEF group. Similarly, ascites was detected via abdominal ultrasound in 5% of HFpEF patients and 20% of HFrEF patients.

IVC measurements were comparable between groups. Patients with HFpEF had a mean IVC diameter of 22.4 mm ± 4.3 mm, with a collapse < 50% in 78% of cases. Those with HFrEF had a mean diameter of 22.8 mm ± 4.8 mm, and 70% exhibited a collapse < 50%.

It is important to highlight that, of the 90 patients included in our study, the VExUS protocol was successfully performed in 80 individuals. In the remaining 10 cases, inadequate acoustic windows made it technically impossible to obtain Doppler waveforms from one or more of the venous structures required for scoring—primarily the intrarenal veins. Since complete evaluation is necessary for accurate classification, no VExUS grade was assigned when any component was missing.

Regarding hepatic vein Doppler assessment, no statistically significant differences were found ($p = 0.242$). Among HFpEF patients, 26% had normal flow, 41% showed moderate alterations, and 33% had severely abnormal flow. In the HFrEF group, these figures were 12%, 58%, and 30%, respectively. Similarly, no significant differences were observed in portal vein Doppler patterns ($p = 0.355$). In the HFpEF group, 35% had normal flow, 43% showed pulsatility < 50%, and 22% had pulsatility > 50%. In HFrEF patients, 24% had normal flow, 43% had pulsatility < 50%, and 43% showed pulsatility > 50%. As for intrarenal venous flow, the distribution was also similar ($p = 0.839$). In HFpEF patients, 55% exhibited continuous flow and 45% discontinuous flow (35% biphasic, 10% monophasic). In the HFrEF group, 63% had continuous flow and 37% discontinuous flow (30% biphasic, 7% monophasic) (Figure 2).

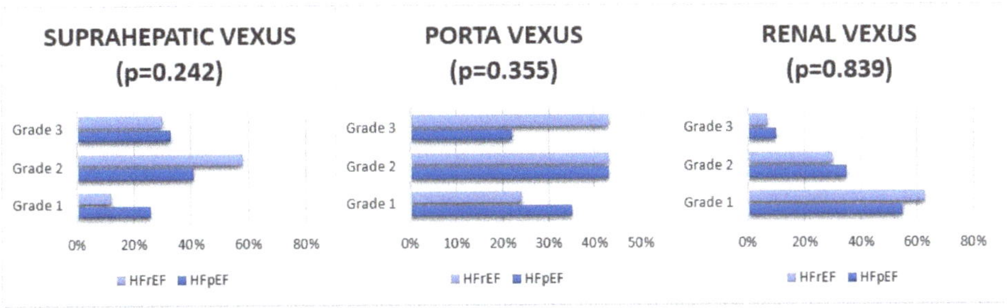

Figure 2. VExUS assessment. Suprahepatic, portal, and renal veins according to LVEF.

According to the VExUS classification, among patients with preserved ejection fraction (HFpEF), 20% were classified as grade 0 (no congestion), 39% as grade 1 (mild congestion), 26% as grade 2 (moderate congestion), and 15% as grade 3 (severe congestion). In the reduced ejection fraction group (HFrEF), 19% were classified as grade 0, 31% as grade 1, 38% as grade 2, and 12% as grade 3 (Figure 3).

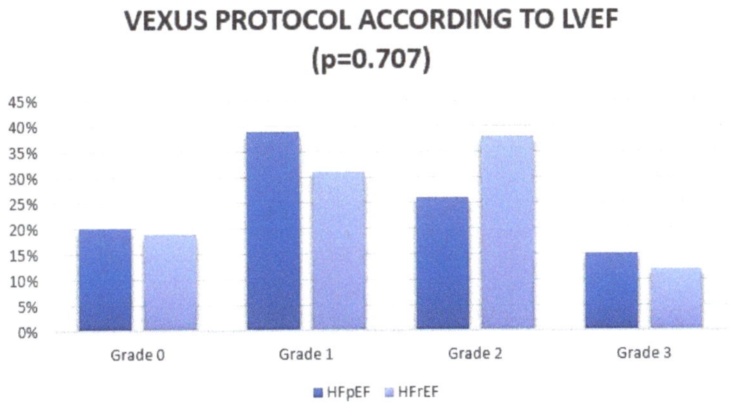

Figure 3. VExUS protocol according to LVEF.

When the cohort was dichotomized into low (VExUS grades 0–1) and moderate-to-severe (VExUS grades 2–3) congestion, 59% of HFpEF patients fell into the low-congestion group and 41% into the higher-congestion group. Among HFrEF patients, the distribution was evenly split (50% vs. 50%).

As previously described, there were no significant differences between the groups in terms of VExUS grading—neither in the four-grade classification ($p = 0.707$) nor in the dichotomized analysis ($p = 0.526$).

NT-proBNP and CA125 levels were significantly higher in HFrEF patients (median 11,278 vs. 4665 ng/L, $p < 0.001$ and 70.6 vs. 37.3 U/mL, $p = 0.032$, respectively). Although not statistically significant, HFrEF patients showed a lower estimated glomerular filtration rate and higher serum creatinine levels. Other parameters, including serum and urinary electrolytes, hemoglobin, and inflammatory markers, showed no significant intergroup differences (Table 1).

Table 1. Clinical, analytical, and ultrasound variables according to LVEF.

Variables	HFpEF (>50%)	HFrEF (<50%)	p-Value
Hypertension	52 (87)	26 (87)	1.000
Diabetes	20 (33)	11 (37)	0.754
Flutter/Atrial fibrillation	52 (87)	16 (53)	<0.001 *
Dyslipidemia	33 (55)	19 (63)	0.451
CKD	17 (28)	15 (50)	0.043 *
Congestion score (points)	4 ± 2	5 ± 2	0.119
Pleural effusion	31 (52)	23 (77)	0.022 *
Ascites	3 (5)	6 (20)	0.020 *
Inferior vena cava distension (mm)	22.4 ± 4.3	22.8 ± 4.8	0.693
VExUS grade 0–1	31 (59)	14 (50)	0.526
VExUS grade 2–3	21 (41)	14 (50)	0.526
Uric acid (mg/dL)	7.0 ± 3.8	7.2 ± 3.5	0.466
Urea (mg/dL)	66.7 ± 34	75.3 ± 23	0.223
Creatinine (mg/dL)	1.3 ± 0.5	1.5 ± 0.5	0.077
eGFR (mL/min/1.73 m^2)	52 ± 29	44 ± 24	0.270
Sodium (mmol/L)	141 ± 3.5	140 ± 5.1	0.097
Potassium (mmol/L)	4.02 ± 0.4	4.31 ± 0.9	0.056
Chloride (mmol/L)	100 ± 5	101 ± 6	0.820
Creatinine in urine (mg/dL)	37.5 ± 28.5	33.7 ± 21.3	0.564
Chloride in urine (mg/dL)	86 ± 59	87 ± 63	0.960
Sodium in urine (mg/dL)	93.5 ± 48.5	88 ± 61	0.438
Potassium in urine (mg/dL)	25.1 ± 10.1	27.4 ± 11.3	0.354
Urea in urine (mg/dL)	522 ± 412	536 ± 470	0.883
ACR in urine (mg/g)	92 (RIC: 237)	121 (RIC: 236)	0.406
NTproBNP (ng/dL)	4665 (RIC: 5662)	11,278 (RIC: 20,179)	<0.001 *
Ca 125 (U/mL)	37.3 (RIC: 51.6)	70.6 (RIC: 120)	0.032 *
Hemoglobin (g/dL)	12.4 ± 2	12 ± 2.4	0.368
Hematocrit (%)	38.2 ± 6.1	36.3 ± 7.5	0.215

Categorical variables are shown as number of patients (n) and percentage of total (%). Units are given for quantitative variables. * Variables with p-values <0.05 indicate significant variation between HF phenotypes.

There were no significant differences in intravenous furosemide doses or the use of hypertonic saline. However, higher VExUS scores correlated with a greater use of intravenous diuretics. At discharge, guideline-directed medical therapy was underutilized, particularly in HFpEF patients.

There were no significant differences in in-hospital mortality (5% HFpEF vs. 10% HFrEF, $p = 0.370$), 3-month cardiovascular mortality (8% vs. 14%, $p = 0.514$), or all-cause readmission (14% vs. 26%, $p = 0.274$). Diuretic rescue was rare in both groups.

4. Discussion

In this single-center cohort study of patients hospitalized with AHF, the quantification of venous congestion at admission using the VExUS protocol showed no significant differences between those with HFpEF and those with HFrEF. Contrary to our initial hypothesis, this suggests that the VExUS score may serve as a reliable tool for assessing systemic congestion across the full spectrum of heart failure phenotypes. These findings are aligned with recent evidence highlighting the utility of multimodal congestion assessment irrespective of LVEF and support the broader clinical applicability of the VExUS protocol.

The baseline characteristics of our cohort mirrored those reported in national and international HF registries [24], with a predominance of HFpEF patients. Atrial fibrillation and anticoagulation therapy were more frequent in this group, in agreement with previous studies [24–26]. CKD was more prevalent in the HFrEF group, despite being typically associated with HFpEF [26]. Severe valvular disease and implantable cardioverter-defibrillators were also more common in HFrEF, reflecting the advanced structural damage characteristic of this phenotype and its specific therapeutic indications [27].

Regarding the degree of congestion assessed by physical examination, no significant differences were found between groups; however, tissue edema appeared more frequently in patients with reduced ejection fraction (HFrEF). This group also showed a higher prevalence of pleural effusion and ascites, likely reflecting volume overload-driven decompensation—typical of HFrEF—even though a mixed mechanism involving both vascular redistribution and volume expansion is increasingly recognized [28].

Van Aelst et al. [29] reported similar findings, noting that both the preserved (HFpEF) and reduced ejection fraction groups showed comparable levels of venous congestion in acute heart failure, despite the higher natriuretic peptide levels seen in HFrEF patients. Consistently, our study found increased levels of both intravascular (NT-proBNP) and tissue congestion markers (CA125) in HFrEF patients. However, recent studies have reported elevated CA125 levels in HFpEF patients compared to controls, suggesting a potential role for CA125 as a marker of congestion and prognosis in HFpEF [30,31]. The discrepancies between our findings and those of these studies may be attributed to differences in sample size, patient characteristics, or methodological approaches. These variations underscore the complexity of heart failure phenotyping and the need for further research to clarify the role of CA125 in HFpEF. Additionally, VExUS, as a direct ultrasound-based tool for assessing venous congestion, may provide complementary information to biomarker profiles, potentially enhancing our understanding of congestion pathways in HFpEF and HFrEF.

A recent meta-analysis by Palazzuoli et al. [32] emphasized the high prevalence of clinical and ultrasound signs of congestion across the heart failure spectrum, highlighting the importance of a multimodal approach to improve risk stratification. Our findings align with this conclusion, showing no differences in sonographic congestion patterns, which reinforces its value in tailoring therapy.

While physical signs of congestion were similarly distributed, pleural effusion and ascites were significantly more frequent in HFrEF, consistent with its greater volume overload tendency [28]. Nonetheless, no differences were observed in VExUS grading, supporting the findings from Van Aelst et al. [29] and a recent meta-analysis by Palazzuoli et al. [32], which emphasized the high prevalence and prognostic value of congestion markers regardless of LVEF. Our data reinforce these conclusions and advocate for VExUS as a complementary tool in bedside evaluation.

Although some subgroup analyses, such as that of Bhardwaj et al. [33], suggest that VExUS-based congestion grading is independent of LVEF, further dedicated studies are needed—underscoring the relevance and future direction of our work.

Diuretic and hypertonic saline use did not differ by LVEF; higher VExUS scores correlated with increased intravenous diuretic requirements. This highlights its potential utility for tailoring depletive therapy. Clinical outcomes such as in-hospital mortality, three-month cardiovascular mortality, and readmissions showed no significant intergroup differences, possibly due to the limited sample size and follow-up duration [29–32,34].

Finally, we identified important gaps in baseline treatment. Many patients were underprescribed guideline-directed medical therapy upon admission despite prior LVEF classification, particularly regarding the use of disease-modifying therapies such as iSGLT2 inhibitors. This is concerning, considering current ESC guidelines which recommend these agents for both HFpEF and HFrEF patients [27]. Recent national data (RICA2 registry) showed a prescription rate of 68% for iSGLT2 inhibitors [35], highlighting a disparity that warrants further exploration.

The strengths of this study include its prospective design and comprehensive congestion assessment using multimodal tools. Limitations include the single-center setting, modest sample size, and the lack of long-term follow-up. Further multicenter studies are warranted to confirm these findings and explore the prognostic value of VExUS in diverse HF populations.

In summary, while HF phenotypes differ in traditional markers of congestion, VExUS provides consistent assessment across groups, supporting its utility as a universal tool for guiding therapy in AHF.

Study Limitations

This study has several limitations that should be acknowledged. First, it is a single-center observational study with a limited sample size, particularly within subgroups such as those with HFrEF, which may reduce statistical power and increase the risk of Type II error. As this was an exploratory study, no formal power calculation was performed. Second, challenges inherent to ultrasonographic interpretation exist: for example, IVC diameter can be elevated in the absence of true congestion due to factors like low body mass index, athletic conditioning, pulmonary hypertension, or severe tricuspid regurgitation. Additionally, other non-congestive factors such as liver disease, anemia, and obesity may also influence VExUS measurements and biomarker levels, potentially confounding their interpretation. Furthermore, while patients with advanced chronic kidney disease (CKD stages 4–5), dialysis dependence, or renal transplant were excluded, individuals with mild-to-moderate CKD (stages 1–3) or subclinical liver dysfunction were included, potentially confounding VExUS interpretation due to altered venous Doppler waveforms independent of volume status.

Ultrasound examinations were performed using two portable devices (Lumify® and Omega®), with Lumify® generally providing lower image quality. However, scans were conducted by trained sonographers using a standardized VExUS protocol, and reproducibility was substantial (Cohen's kappa 0.80), supporting consistent scoring despite device differences. Some variability related to device differences cannot be excluded.

Given these factors, the ultrasound findings must be interpreted cautiously within the broader clinical context, underscoring the importance of individualized assessment when guiding decongestive therapy. Future multicenter studies with larger samples are warranted to validate these findings and explore their prognostic significance.

5. Conclusions

In this study, no significant differences were found in the ultrasonographic assessment of venous congestion using the VExUS protocol between patients with acute heart failure (AHF) and preserved (HFpEF) versus reduced ejection fraction (HFrEF), despite notable differences in biomarkers and clinical signs of congestion. These findings support the applicability of the VExUS protocol as an objective and standardized tool for evaluating systemic congestion in AHF patients, regardless of LVEF phenotype. Its implementation may contribute to improved risk stratification and a more precise tailoring of decongestive therapy.

VExUS may represent a valuable addition to the clinical toolkit for the personalized management of acute heart failure.

Supplementary Materials: The following supporting information can be downloaded at https://www.mdpi.com/article/10.3390/app15158157/s1, VExUS Scoring Reproducibility analysis.

Author Contributions: Conceptualization, A.A.-O. and J.R.-G.; Data curation, L.K.E.-S., A.C.-S.d.S., Z.S.A.F. and S.C.-A.; Formal analysis, A.C.-S.d.S., M.S.-M., V.G.-H. and A.A.-O.; Investigation, L.K.E.-S., A.C.-S.d.S., Z.S.A.F. and S.C.-A.; Resources, M.S.-M., V.G.-H., A.A.-O. and I.G.-L.; Software, A.A.-O.; Supervision, J.R.-G.; Visualization, L.K.E.-S.; Writing—original draft, L.K.E.-S.; Writing—review and editing, I.G.-L. and J.R.-G. All authors have read and agreed to the published version of the manuscript.

Funding: Project PID2022-139143OA-I00 was funded by MICIU/AEI/10.13039/501100011033 and by ERDF/EU.

Institutional Review Board Statement: This study was conducted in accordance with the Declaration of Helsinki and approved by the Ethics Committee of Aragón (CEICA) (protocol code COL21/003; date of approval: February 2023).

Informed Consent Statement: Informed consent was obtained from all subjects involved in this study.

Data Availability Statement: The data that support the findings of this study are available from the corresponding author upon reasonable request. Due to privacy and ethical restrictions, raw data are not publicly available.

Acknowledgments: The authors thank the Internal Medicine Department of Hospital Clínico Universitario Lozano Blesa for their support and the patients who participated in this study.

Conflicts of Interest: The authors declare no conflicts of interest.

Abbreviations

The following abbreviations are used in this manuscript:

HF	Heart failure
LVEF	Left ventricular ejection fraction
VExUS	Venous Excess Ultrasound
HFpEF	Heart failure with preserved ejection fraction
HFrEF	Heart failure with reduced ejection fraction
CA125	Carbohydrate antigen 125
NTproBNP	N-terminal pro-B-type natriuretic peptide
POCUS	Point-of-care ultrasonography
LUS	Lung ultrasound
IVC	Inferior vena cava
AHF	Acute heart failure
CKD	Chronic kidney disease
NYHA	New York Heart Association

References

1. Riccardi, M.; Sammartino, A.M.; Piepoli, M.; Adamo, M.; Pagnesi, M.; Rosano, G.; Metra, M.; von Haehling, S.; Tomasoni, D. Heart failure: An update from the last years and a look at the near future. *ESC Heart Fail.* **2022**, *9*, 3667–3693. [CrossRef] [PubMed]
2. Husain-Syed, F.; McCullough, P.A.; Birk, H.W.; Renker, M.; Brocca, A.; Seeger, W.; Ronco, C. Cardio-Pulmonary-Renal Interactions: A Multidisciplinary Approach. *J. Am. Coll. Cardiol.* **2015**, *65*, 2433–2448. [CrossRef] [PubMed]
3. Sicras-Mainar, A.; Sicras-Navarro, A.; Palacios, B.; Varela, L.; Delgado, J.F. Epidemiology and treatment of heart failure in Spain: The HF-PATHWAYS study. *Rev. Esp. Cardiol.* **2022**, *75*, 31–38. [CrossRef] [PubMed]
4. Kazory, A.; Elkayam, U. Cardiorenal interactions in acute decompensated heart failure: Contemporary concepts facing emerging controversies. *J. Card. Fail.* **2014**, *20*, 1004–1011. [CrossRef] [PubMed]
5. Sánchez-Marteles, M.; Garcés-Horna, V.; Morales-Rull, J.L.; Casado, J.; Carrera-Izquierdo, M.; Conde-Martel, A.; Dávila-Ramos, M.F.; Llácer, P.; Salamanca-Bautista, P.; Ruiz, R.; et al. Combining loop and thiazide diuretics across the left ventricular ejection fraction spectrum. *JACC Heart Fail.* **2024**, *12*, 1719–1730. [CrossRef] [PubMed]
6. Koratala, A.; Kazory, A. Point of Care Ultrasonography for Objective Assessment of Heart Failure: Integration of Cardiac, Vascular, and Extravascular Determinants of Volume Status. *Cardiorenal Med.* **2021**, *11*, 5–17. [CrossRef] [PubMed]
7. Rubio-Gracia, J.; Demissei, B.G.; ter Maaten, J.M.; Cleland, J.G.; O'Connor, C.M.; Metra, M.; Ponikowski, P.; Teerlink, J.R.; Cotter, G.; Davison, B.A.; et al. Prevalence, predictors and clinical outcome of residual congestion in acute decompensated heart failure. *Int. J. Cardiol.* **2018**, *258*, 185–191. [CrossRef] [PubMed]
8. Morrow, D.A.; Velazquez, E.J.; Devore, A.D.; Desai, A.S.; Duffy, C.I.; Ambrosy, A.P.; Gurmu, Y.; Mccague, K.; Rocha, R.; Braunwald, E. Clinical Outcomes in Patients with Acute Decompensated Heart Failure Randomly Assigned to Sacubitril/Valsartan or Enalapril in the PIONEER-HF Trial. *Circulation* **2019**, *139*, 2285–2288. [CrossRef] [PubMed]
9. Long, B.; Koyfman, A.; Gottlieb, M. Diagnosis of acute heart failure in the emergency department: An evidence-based review. *West. J. Emerg. Med.* **2019**, *20*, 875–884. [CrossRef] [PubMed]
10. Maisel, A.; Mueller, C.; Adams, K., Jr.; Anker, S.D.; Aspromonte, N.; Cleland, J.G.; Cohen-Solal, A.; Dahlstrom, U.; DeMaria, A.; Di Somma, S.; et al. State of the art: Using natriuretic peptide levels in clinical practice. *Eur. J. Heart Fail.* **2008**, *10*, 824–839. [CrossRef] [PubMed]
11. Koratala, A.; Reisinger, N. POCUS for nephrologists: Basic principles and a general approach. *Kidney360* **2021**, *2*, 1660–1668. [CrossRef] [PubMed]
12. Jobs, A.; Vonthein, R.; König, I.R.; Schäfer, J.; Nauck, M.; Haag, S.; Fichera, C.F.; Stiermaier, T.; Ledwoch, J.; Schneider, A.; et al. Inferior vena cava ultrasound in acute decompensated heart failure: Design rationale of the CAVA-ADHF-DZHK10 trial. *ESC Heart Fail.* **2020**, *7*, 973–983. [CrossRef] [PubMed]
13. Rola, P.; Miralles-Aguiar, F.; Argaiz, E.; Beaubien-Souligny, W.; Haycock, K.; Karimov, T.; Dinh, V.A.; Spiegel, R. Clinical applications of the venous excess ultrasound (VExUS) score: Conceptual review and case series. *Ultrasound J.* **2021**, *13*, 32. [CrossRef] [PubMed]
14. Pastore, M.C.; Ilardi, F.; Stefanini, A.; Mandoli, G.E.; Palermi, S.; Bandera, F.; Benfari, G.; Esposito, R.; Lisi, M.; Pasquini, A.; et al. Bedside ultrasound for hemodynamic monitoring in cardiac intensive care unit. *J. Clin. Med.* **2022**, *11*, 7538. [CrossRef] [PubMed]
15. Narula, J.; Chandrashekhar, Y.; Braunwald, E. Time to Add a Fifth Pillar to Bedside Physical Examination: Inspection, Palpation, Percussion, Auscultation, and Insonation. *JAMA Cardiol.* **2018**, *3*, 346–350. [CrossRef] [PubMed]
16. Llàcer, P.; Romero, G.; Trullàs, J.C.; de la Espriella, R.; Cobo, M.; Quiroga, B.; Casado, J.; Slon-Roblero, M.F.; Morales-Rull, J.L.; Morgado, J.I.; et al. Consensus on the approach to hydrosaline overload in acute heart failure. SEMI/SEC/S.E.N. recommendations. *Rev. Esp. Cardiol.* **2024**, *77*, 556–565. [CrossRef] [PubMed]
17. Beaubien-Souligny, W.; Rola, P.; Haycock, K.; Bouchard, J.; Lamarche, Y.; Spiegel, R.; Denault, A.Y. Quantifying systemic congestion with Point-Of-Care ultrasound: Development of the venous excess ultrasound grading system. *Ultrasound J.* **2020**, *12*, 16. [CrossRef] [PubMed]
18. Assavapokee, T.; Rola, P.; Assavapokee, N.; Koratala, A. Decoding VExUS: A practical guide for excelling in point-of-care ultrasound assessment of venous congestion. *Ultrasound J.* **2024**, *16*, 48. [CrossRef] [PubMed]
19. Landi, I.; Guerritore, L.; Iannaccone, A.; Ricotti, A.; Rola, P.; Garrone, M. Assessment of venous congestion with venous excess ultrasound score in the prognosis of acute heart failure in the emergency department: A prospective study. *Eur. Heart J. Open.* **2024**, *4*, oeae050. [CrossRef] [PubMed]
20. Gamarra, Á.; Salamanca, J.; Díez-Villanueva, P.; Cuenca, S.; Vázquez, J.; Aguilar, R.J.; Diego, G.; Rodríguez, A.P.; Alfonso, F. Ultrasound imaging of congestion in heart failure: A narrative review. *Cardiovasc Diagn Ther.* **2025**, *15*, 233–250. [CrossRef] [PubMed]
21. Rubio Gracia, J.; Sánchez Marteles, M.; Pérez Calvo, J.I. Involvement of systemic venous congestion in heart failure. *Rev. Clin. Esp.* **2017**, *217*, 161–169. [CrossRef] [PubMed]

22. Di Fiore, V.; Del Punta, L.; De Biase, N.; Pellicori, P.; Gargani, L.; Dini, F.L.; Armenia, S.; Vigni, M.L.; Maremmani, D.; Masi, S. Integrative assessment of congestion in heart failure using ultrasound imaging. *Int. Emerg. Med.* **2025**, *20*, 11–22. [CrossRef] [PubMed]
23. McMurray, J.J.V.; Packer, M.; Desai, A.S.; Gong, J.; Lefkowitz, M.P.; Rizkala, A.R.; Rouleau, J.L.; Shi, V.C.; Solomon, S.D.; Swedberg, K.; et al. Angiotensin–Neprilysin Inhibition versus Enalapril in Heart Failure. *N. Engl. J. Med.* **2014**, *371*, 993–1004. [CrossRef] [PubMed]
24. Omote, K.; Verbrugge, F.H.; Borlaug, B.A. Heart failure with preserved ejection fraction: Mechanisms and treatment strategies. *Annu. Rev. Med.* **2022**, *73*, 321–337. [CrossRef] [PubMed]
25. Borlaug, B.A.; Sharma, K.; Shah, S.J.; Ho, J.E. Heart failure with preserved ejection fraction: JACC Scientific Statement. *J. Am. Coll. Cardiol.* **2023**, *81*, 1810–1834. [CrossRef] [PubMed]
26. Kondo, T.; Dewan, P.; Anand, I.S.; Desai, A.S.; Packer, M.; Ziler, M.R.; Pfeffer, M.A.; Solomon, S.D.; Abraham, W.T.; Shah, S.J.; et al. Clinical characteristics and outcomes in patients with heart failure: Are there thresholds justifying a clínical classification? *Circulation* **2023**, *148*, 732–749. [CrossRef] [PubMed]
27. McDonagh, T.A.; Metra, M.; Adamo, M.; Gardner, R.S.; Baumbach, A.; Bohm, M.; Burri, H.; Butler, J.; Čelutkienė, J.; Chioncel, O.; et al. 2021 ESC Guidelines for the diagnosis and treatment of acute and chronic heart failure: Developed by the Task Force for the diagnosis and treatment of acute and chronic heart failure of the European Society of Cardiology (ESC) With the special contribution of the Heart Failure Association (HFA) of the ESC. *Rev. Esp. Cardiol.* **2022**, *75*, 523. [PubMed]
28. Espriella, R.; Santas, E.; Zegri Reiriz, I.; Górriz, J.L.; Cobo Marcos, M.; Nuñez, J. Quantification and treatment of congestion in heart failure: A clinical and pathophysiological overview. *Nefrología* **2022**, *42*, 145–162. [CrossRef] [PubMed]
29. Van Aelst, L.N.L.; Arrigo, M.; Placido, R.; Akiyama, E.; Girerd, N.; Zannad, F.; Manivet, P.; Rossignol, P.; Badoz, M.; Sadoune, M.; et al. Acutely decompensated heart failure with preserved and reduced ejection fraction present with comparable haemodynamic congestion. *Eur. J. Heart Fail.* **2018**, *20*, 738–747. [CrossRef] [PubMed]
30. Menghoum, N.; Badii, M.C.; Deltombe, M.; Lejeune, S.; Roy, C.; Vancraeynest, D.; Pasquet, A.; Gerber, B.L.; Horman, S.; Gruson, D.; et al. Carbohydrate antigen 125: A useful marker of congestion, fibrosis, and prognosis in heart failure with preserved ejection fraction. *ESC Heart Fail.* **2024**, *11*, 1493–1505. [CrossRef] [PubMed]
31. Tuersun, R.; Abudouwayiti, A.; Li, Y.; Pan, Y.; Aimaier, S.; Wen, Z.Y.; Gao, W.T.; Ma, L.J.; Mahemuti, A.; Zheng, Y.Y. Serum CA125: A prognostic biomarker for mortality in chronic heart failure. *BMC Cardiovasc. Disord.* **2025**, *25*, 227. [CrossRef] [PubMed]
32. Palazzuoli, A.; Ruocco, G.; Pellicori, P.; Gargani, L.; Coiro, S.; Lamiral, Z.; Ambrosio, G.; Rastogi, T.; Girerd, N. Multi-modality assessment of congestion in acute heart failure: Associations with left ventricular ejection fraction and prognosis. *Curr. Probl. Cardiol.* **2024**, *49*, 102374. [CrossRef] [PubMed]
33. Bhardwaj, V.; Vikneswaran, G.; Rola, P.; Raju, S.; Bhat, R.S.; Jayakumar, A.; Alva, A. Combination of Inferior Vena Cava Diameter, Hepatic Venous Flow, and Portal Vein Pulsatility Index: Venous Excess Ultrasound Score (VEXUS Score) in Predicting Acute Kidney Injury in Patients with Cardiorenal Syndrome: A Prospective Cohort Study. *Indian J. Crit. Care Med.* **2020**, *9*, 7839. [CrossRef] [PubMed]
34. Núñez-Ramos, J.A.; Aguirre-Acevedo, D.C.; Pana-Toloza, M.C. Point of care ultrasound impact in acute heart failure hospitalization: A retrospective cohort study. *Am. J. Emerg. Med.* **2023**, *66*, 141–145. [CrossRef] [PubMed]
35. Trullàs, J.C.; Moreno-García, M.C.; Mittelbrunn-Alquézar, V.; Conde-Martel, A.; Soler-Rangel, L.; Montero-Pérez-Barquero, M.; Casado, J.; Sánchez-Marteles, M.; Arévalo-Lorido, J.C.; Pérez-Silvestre, J.; et al. The RICA-2 registry: Design and baseline characteristics of the first 1000 patients. *Rev. Clin. Esp.* **2024**, *224*, 522–533. [CrossRef] [PubMed]

applied sciences

MDPI

Article

The Role of Ultrasonography in the Diagnostic Evaluation of Patients with Polyneuropathy

Maj Jožef and Simon Podnar *

Institute of Clinical Neurophysiology, Division of Neurology, University Medical Center Ljubljana, 1525 Ljubljana, Slovenia; majjozef@gmail.com
* Correspondence: simon.podnar@kclj.si

Abstract: Background: The role of ultrasonography (US) in the practical management of polyneuropathies, particularly axonal, remains unclear. The present study aimed to explore the contribution of the US examination of polyneuropathies in daily clinical practice. Methods: We performed a retrospective chart review of patients with clinical and electrophysiological diagnoses of polyneuropathy referred to our US laboratory over eight years. The contribution of US examination in this patient population was evaluated. Results: We analyzed 201 consecutive patients (66% men), aged 12–90 years (mean (SD), 62 (15) years). The most common referral questions were differentiation of hereditary from acquired demyelinating polyneuropathies (71 (35%) patients, sensitivity 63%, specificity 88%), and additional focal neuropathies in patients with generalized neuropathies (51 (25%) patients, sensitivity 75%, specificity 34%). The US examination was pathological in 158 (79%) of patients. The most common US finding was nerve enlargement at typical entrapment sites (73 (36%) patients), followed by proximal nerve thickening (34 (17%) patients). The US provided new diagnoses in 7 (3.5%) patients, contributed to diagnoses in 39 (19%) patients, and confirmed diagnoses in 50 (25%) patients. Conclusion: Our study demonstrated the ability of peripheral nerve US to provide useful additional diagnostic information in about half of the referred patients with polyneuropathy.

Keywords: electrodiagnosis; nerve cross-sectional area; peripheral nerves; polyneuropathy; ultrasonography

1. Introduction

Clinical features supported by nerve conduction studies (NCS) and needle electromyography (EMG) remain a cornerstone in managing polyneuropathies. However, clinical and NCS criteria are complex and invasive, focusing primarily on specificity. For example, NCSs require limb temperature control, whereas needle EMG requires patient tolerance and is subject to sampling bias. Misinterpretation and misdiagnosis are common, resulting in the exclusion of patients who could benefit from treatment, for example, in chronic inflammatory demyelinating polyneuropathy (CIDP) and multifocal motor neuropathy (MMN). Therefore, more sensitive, practical, and non-invasive diagnostic tools are needed in clinical settings.

Nerve ultrasound (US) is a relatively novel non-invasive diagnostic tool for evaluating nerves, particularly superficial ones. Nerves are recognized by hypoechogenic fascicles, encircled by hyperechogenic fibrous tissue, i.e., a honeycomb structure. The most relevant US features are echogenicity and nerve size. The latter is assessed in the transverse plane and expressed as the cross-sectional area (CSA) in mm^2 [1].

With advancements in US technology, a detailed localization, morphological, and etiological evaluation of peripheral nerves has become possible [1,2]. Initially, the complementary role of nerve US was recognized in focal compression and entrapment mononeuropathies [3]. In clinical practice, US is commonly used in the diagnosis of carpal tunnel syndrome (CTS) and ulnar neuropathy (UNE). Later, the diagnostic potential of nerve US was also shown in polyneuropathies, particularly demyelinating [4], either immune-mediated [5], hereditary [6], or infectious [5]. Typical US findings in polyneuropathies include marked uniform nerve enlargement in Charcot-Marie-Tooth disease type 1A (CMT1A), whereas other CMT types have mild uniform nerve enlargement, more pronounced at entrapment sites. In chronic inflammatory demyelinating polyneuropathy (CIDP), nerve enlargement is less pronounced and uniform, expressed particularly in proximal nerve segments [7]. By contrast, segmental nerve enlargement is typical of multifocal motor neuropathy (MMN) [8] and multifocal CIDP [9]. In contrast, the most common axonal polyneuropathies usually have normal US findings or mild to moderate nerve enlargement [10].

Nerve US is useful in Guillain-Barré syndrome (GBS) showing particularly proximal nerve enlargement in the acute phase, 1–3 days after the onset of symptoms and before the appearance of NCS abnormalities [11]. Nerve enlargement may persist after the symptoms are resolved, or nerves may even become thinner. The US can also help to distinguish MMN from amyotrophic lateral sclerosis (ALS) by detecting multifocal nerve enlargements in the former [12].

Several US protocols and scoring systems have been proposed for use in polyneuropathies, ranging from short and simple [13] to more extensive and complex [14]. The former includes the quantification of the pattern and extent of nerve enlargement, such as intra- and inter-nerve variability between two nerves, while the latter includes the location and the degree of nerve enlargement between both sides as well.

In addition to its diagnostic value, nerve US also has therapeutic applications. Low Intensity Pulsed Ultrasound (LIPUS) promotes nerve regeneration by stimulating Schwann cells and axonal growth, particularly in sciatic or facial nerve injuries. It is also used for neuromodulation in the treatment of neuropathic pain [15]. High Intensity Focused Ultrasound (HIFUS) is used for non-invasive nerve ablation in the treatment of chronic pain (trigeminal neuralgia, Morton's neuroma) using thermal and mechanical signals to disrupt pain transmission [16]. Phonophoresis is used to enhance the transdermal delivery of medications (corticosteroids, lidocaine) in the treatment of neural inflammation [17]. However, the actual contribution of the US in the daily diagnosis and management of patients with polyneuropathies has been only rarely studied [18]. Therefore, the position and indications for US studies in polyneuropathy evaluation protocols are unclear. Nevertheless, the US is gaining recognition as demonstrated by the inclusion of nerve US criteria in the last edition of the European Academy of Neurology Guidelines for diagnosing and treating CIDP [19].

We performed a retrospective chart review of a series of consecutive polyneuropathy patients referred to our neuro-muscular US unit. The primary outcome of our study was the nerve US findings and sensitivity in patients with various polyneuropathies. The secondary outcome was the contribution of nerve US in the diagnosis of common polyneuropathies.

2. Materials and Methods

2.1. Study Design

Our retrospective single-center study included patients referred to the US laboratory at the Institute of Clinical Neurophysiology, University Medical Center Ljubljana, Slovenia from December 2012 to August 2021. Our US unit patients are mainly referred following electrodiagnostic (EDx) examination. We obtained clinical and EDx information from patients' electronic medical charts. We included patients with ICD-10 diagnoses G60–G64 (i.e., polyneuropathies and other disorders of the peripheral nervous system) and divided them according to their final polyneuropathy diagnoses. Patients referred to as mononeuropathy with additional US findings (e.g., nerve thickening at non-entrapment sites) were further evaluated at our institution. The diagnostic contribution of the US was determined as one of four categories.

2.2. Study Population

We included 209 patients with final diagnosis of polyneuropathy, but excluded 7 patients with motor neuron disease and 1 patient with neurofibromatosis.

2.3. Measures and Evaluation

From the electronic medical charts, we extracted patients' main demographic and clinical features, including age, sex, EDx, and US findings.

2.4. Primary Outcome Measures

2.4.1. Electrodiagnostic Evaluation (EDx)

EDx examination was performed according to a standard polyneuropathy protocol that includes motor nerve conduction studies (NCS) of the median, ulnar, fibular, and tibial nerves, as well as sensory NCSs of the median, ulnar, and sural nerves, at least unilaterally. Measured NCS parameters included motor and sensory latency, F-wave latency, response duration, response amplitude and area, temporal dispersion, conduction block (at least 30% amplitude reduction), and motor and sensory conduction velocities. These parameters were compared to our laboratory reference values (Table 1). The evaluation protocol also included a needle electromyography (EMG) examination of at least two leg muscles to assess spontaneous denervation activity or reinnervation changes. Based on clinical and EDx criteria for the type of polyneuropathy (Table 2), patients' diagnoses were divided into: (1) demyelinating polyneuropathy, (2) axonal polyneuropathy, (3) mixed polyneuropathy, and (4) multifocal neuropathy. Several differential diagnostic questions were formulated: (1) demyelinating vs. axonal polyneuropathy, (2) hereditary vs. acquired polyneuropathy, (3) demyelinating vs. diabetic polyneuropathy, (4) MMN vs. motor neuron disease, (5) focal neuropathy superimposed on polyneuropathy, (6) multiple mononeuropathy vs. polyneuropathy, and (7) diagnostic question not clear.

Table 1. EDx reference values for the main motor and sensory nerves used in the present study.

Nerve	Latency (ms)	Amplitude (mV/uV)	Velocity (m/s)	F-Wave Latency (ms)
Motor				
Median	<4.4	>4.0	>49	<31
Ulnar	<3.3	>6.0	>49	<32
Radial	<2.9	>2.0	>49	
Fibular	<6.5	>2.0	>44	<56
Tibial	<5.8	>4.0	>41	<56
Sensory				
Median		>20	>50	
Ulnar		>17	>50	
Radial		>15	>50	
Sural		>6	>40	
Medial plantar		>2	>35	

Table 2. Electrodiagnostic (EDx) criteria for different types of polyneuropathies evaluated in the study.

Polyneuropathy	EDx Criteria
Demyelinating	- Distal motor latency ≥ 50% above ULN in 2 nerves - Motor conduction velocity < 70% of LLN in 2 nerves - F-wave latency ≥ 20% above ULN in 2 nerves - Motor conduction block ≥ 30% - Abnormal CMAP temporal dispersion: >30% duration increase - Sensory nerve conduction velocity < 80% of LLN
Axonal	- CMAP amplitude < 80% of LLN in 2 nerves - Normal or slightly reduced conduction velocity - SNAP amplitude < 80% below LLN in 2 nerves - Reduced motor unit recruitment on needle EMG - Spontaneous denervation activity on needle EMG
Mixed	- Combined features of axonal and demyelinating polyneuropathy
Multifocal	- Motor conduction block ≥ 30% at non-entrapment sites in >2 nerves - Asymmetric segmental demyelination - Chronic neurogenic changes on needle EMG

CMAP—Compound Muscle Action Potential; EDx—electrodiagnostic, EMG—electromyography; LLN—the lower limit of normal; SNAP—Sensory Nerve Action Potential; ULN—the upper limit of normal.

2.4.2. Ultrasonographic Evaluation (US)

All patients were US examined by one of the authors (SP) or a neurophysiological technician, both with >5 years of experience in performing US studies of peripheral nerves at the beginning of the study period. At the time of evaluation, US practitioners were not blinded to findings of clinical and EDx examinations. We used standard US equipment (ProSound Alpha 7), and a 4–13 MHz linear array transducer (Hitachi Aloka Medical, Ltd., Tokyo, Japan).

Our main diagnostic criterion was an increase in the nerve cross-sectional area (CSA) measured by manual tracking of the nerve within the hyperechoic rim of the epineurium while positioning the probe perpendicular to the nerve. Our protocol included US evaluation at standard positions: median and ulnar nerves (at the wrist, forearm, elbow, and upper arm), fibular (at the fibular head and in the popliteal fossa), tibial (at the ankle and in the popliteal fossa) and sural nerve (slightly proximal to the ankle). We used a set of our previously reported reference values for nerve CSA [20] (Table 3). When appropriate, we also described changes in nerve echogenicity, thickening of individual nerve fascicles, segmental nerve thickening, and relevant pathologic structures outside of the peripheral nerves. US impressions were described as: (1) generalized nerve enlargement, (2) proximal nerve enlargements, (3) abnormally thin nerves, (4) nerve thickening in proximal limbs, (5) multifocal nerve enlargement at the non-entrapment sites, (6) multifocal nerve enlargement at the entrapment sites, and (7) normal nerves.

Table 3. The upper reference values for the cross-sectional area (CSA) of the main limb nerves used in the present study.

Nerve and Site	CSA (mm^2)
Median	
Wrist	<12
Forearm	<10
Antecubital fossa	<12
Upper arm	<13

Table 3. *Cont.*

Nerve and Site	CSA (mm²)
Ulnar	
Wrist	<8
Forearm	<9
Elbow	<11
Upper arm	<9
Fibular	
Fibular head	<14
Popliteal fossa	<9
Tibial	
Ankle	<18
Popliteal fossa	<39
Sciatic	<46
Sural	<4

We established the contribution of US examination, based on the US criteria for the type of polyneuropathy (Table 4), to the evaluation of patients as one of the following four categories: (1) US revealed new previously unknown information important for diagnosis; (2) US contributive: enhancing clinical and EDx diagnosis by providing relevant additional information; (3) US confirmative: providing evidence confirming clinical and EDx diagnosis; or (4) US not revealing relevant additional information [21].

Table 4. US criteria for the type of polyneuropathy.

Polyneuropathy		US Criteria
Diabetic	-	Variable, often mild-moderate diffuse nerve enlargements (2.5 SD CSA) pronounced in median and tibial nerves
Hereditary	-	Uniform symmetric nerve enlargements (>2.5 SD CSA)
Idiopathic	-	Variable, often mild-moderate diffuse nerve enlargements (1.5 SD CSA)
Typical CIDP	-	Uniform often symmetric nerve enlargement (>2.5 SD CSA)
DADS	-	Uniform symmetric nerve enlargement (>2.5 SD CSA)
Multifocal CIDP	-	Multifocal asymmetric nerve enlargements (>2.0 SD CSA)
MMN	-	Multifocal or segmental nerve enlargements at non-entrapment sites in at least one nerve
Multiple mononeuropathy	-	Moderate asymmetric nerve enlargements (>1.5 SD CSA)
Paraproteinemic polyneuropathy	-	Mild, often asymmetric nerve enlargements (>1.0 SD CSA)
Amyloid polyneuropathy	-	Marked nerve thickening with hypoechoic appearance (>2.5 SD CSA)

2.5. Statistical Analysis

Descriptive statistics were used for the analysis of demographic data. Counts and percentages were provided for categorical variables. Means and standard deviations were calculated for numerical variables. We calculated the sensitivities of US examination in clinical and EDx-confirmed polyneuropathies. Sensitivity was defined as the percentage of patients with clinical and EDx diagnoses of polyneuropathy having pathological US findings compatible with a diagnosis of polyneuropathy.

2.6. Ethic Statements

Due to the retrospective analysis with no direct involvement of patients and data anonymization, the National Ethics Committee of Slovenia waived the necessity of obtaining written informed consent from patients. All analyses were performed by the authors, who were observant to protect all patients' personal information.

3. Results

3.1. Study Population

During the analyzed period, 201 patients with different polyneuropathies who were 12–90 years old met the inclusion criteria (Table 5).

Table 5. Demographics and sensitivity of peripheral nerve ultrasonography (US) in a series of 201 consecutive patients with various polyneuropathies.

Diagnosis	Number (%)	Mean Age (y) (SD)	Males (%)	US Sensitivity (%)
Final cohort	201 (100%)	62 (15)	66	64
Diabetic polyneuropathy	38 (19%)	66 (12)	82	60
Hereditary polyneuropathy	35 (17%)	52 (17)	49	63
Idiopathic polyneuropathy	28 (14%)	67 (14)	75	28
Typical CIDP	22 (11%)	62 (19)	41	82
DADS	12 (6%)	74 (6)	92	58
Multifocal CIDP	11 (5.5%)	56 (12)	91	91
MMN	8 (4%)	62 (14)	50	63
Multiple mononeuropathy	7 (3.5%)	60 (20)	29	57
Paraproteinemic polyneuropathy	5 (2.5%)	70 (6)	80	60
TTR amyloid polyneuropathy	4 (2%)	60 (7)	50	75

CIDP—chronic inflammatory demyelinating polyneuropathy; DADS—distal acquired demyelinating symmetric polyneuropathy; MMN—multifocal motor neuropathy; TTR—transthyretin.

3.2. Electrodiagnostic (EDx) Evaluation

EDx studies were abnormal in 198 (97%) of our 201 patients with EDx reports available at the time of the US study. The most common EDx diagnoses were axonal polyneuropathy (87 patients, 43%), followed by demyelinating (68 patients, 34%) and mixed polyneuropathy (23 patients, 11%; Figure 1). The most common final etiological diagnosis was diabetic polyneuropathy, followed by hereditary polyneuropathy (Table 5).

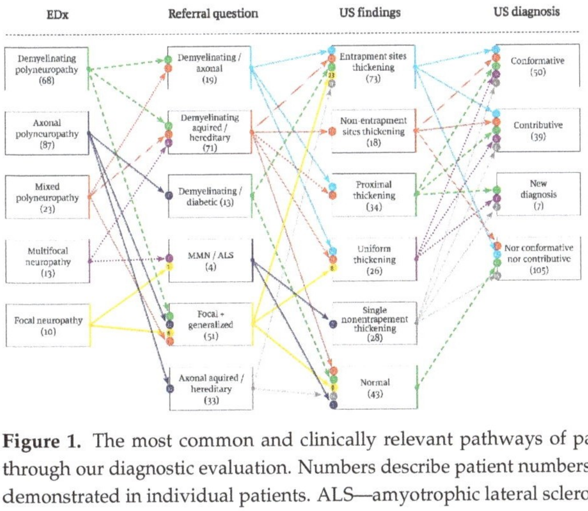

Figure 1. The most common and clinically relevant pathways of patients with polyneuropathy through our diagnostic evaluation. Numbers describe patient numbers. Several US findings may be demonstrated in individual patients. ALS—amyotrophic lateral sclerosis; EDx—electrodiagnostics; MMN—multifocal motor neuropathy; US—ultrasonography.

3.3. Ultrasonography

US examination was abnormal in 158 (79%) of our patients but varied according to the etiology of polyneuropathy and referral questions. Figure 1 shows the frequency of different US findings in our patient population. The most common referral questions for US examination were differentiation of hereditary from acquired demyelinating polyneuropathies (71 patients) and demonstration of additional focal neuropathies in patients with generalized polyneuropathies (51 patients; Figure 1). The sensitivity and specificity for distinguishing hereditary from acquired polyneuropathies were 63% and 88%, respectively, while demonstration of additional focal neuropathies had a sensitivity of 75% and a specificity of 34%. The US provided a new diagnosis in 7 (3%) patients (Table 6), was contributive in 39 (19%) patients, and confirmatory in 50 (25%) of the included patients, but did not provide additional diagnostic information in 105 (52%) patients.

Table 6. Findings in 7 patients with new diagnoses revealed by peripheral nerve ultrasonography (US).

Findings	Number	Referral Diagnosis/Comment
Demyelinating polyneuropathy	1	tarsal tunnel syndrome
Expansive lesions	2	hereditary polyneuropathy
Multifocal CIDP	2	typical CIDP and MGUS
Thoracic outlet syndrome	1	radiation plexopathy
Proximal nerve enlargement	1	small fiber neuropathy

CIDP—chronic inflammatory demyelinating polyneuropathy; MGUS—monoclonal gammopathy of undetermined significance.

3.4. Relation of US to EDx

In our cohort, the most common EDx diagnosis was axonal polyneuropathy (Figure 1), with the most common referral question of superimposed focal neuropathies (35 patients). The most common US finding in this referral group was nerve thickening at the entrapment sites (16 patients), followed by normal findings (8 patients), single focal thickening at the non-entrapment site (5 patients), uniform nerve thickening (5 patients), and proximal nerve thickening (2 patients). Another common referral question in this group was a differentiation of hereditary from acquired polyneuropathy (32 patients), and we most often found normal findings (14 patients), followed by nerve thickening on entrapment sites (10 patients).

Patients with EDx diagnosis of demyelinating polyneuropathies were the second most common, with referral physicians most often trying to differentiate hereditary from acquired forms (59 patients). In this referral group, we also most often found nerve thickening at entrapment sites (20 patients), followed by proximal nerve thickening (13 patients), uniform nerve thickening (11 patients), and nerve thickening at non-entrapment sites (10 patients).

A much less common group of referrals were patients with EDx mixed polyneuropathies, and in these patients, referral physicians most often looked for differentiation between demyelinating and axonal forms (10 patients). In this group, we found proximal nerve thickening (3 patients) and uniform thickening typical for hereditary demyelinating polyneuropathies (1 patient). Another common referral question in this population was to differentiate hereditary from acquired demyelinating polyneuropathy (8 patients).

In patients with EDx multifocal neuropathies, referral physicians tried to differentiate hereditary from acquired demyelinating polyneuropathy (4 patients) and MMN from ALS (1 patient).

4. Discussion

In this retrospective chart review, we explored the role of the US in the diagnostic evaluation of patients with EDx-confirmed polyneuropathy. Most patients in our polyneuropathy cohort were referred to the US to improve their diagnostic formulation as obtained by clinical and EDx examination. US complements the EDx examination by providing precise anatomical location and information on the etiology of the nerve lesion, thereby increasing sensitivity. The US was first used in routine clinical practice for diagnosing focal mononeuropathies, such as median entrapment neuropathy at the wrist (i.e., carpal tunnel syndrome) and ulnar neuropathy. In the former, it is also used to guide therapeutic injections. In the latter, it is important in differentiating the cubital tunnel entrapment and the retrocondylar compression, which determines treatment [22,23]. Later, it became clear that the US can also help in diagnosing polyneuropathies, particularly demyelinating, and the detection of focal neuropathies superimposed on generalized polyneuropathies. In our population, with a large majority (98%) of patients with abnormal EDx findings, the sensitivity of the US examination was moderate (64%). Sensitivity was found to be higher in patients with typical and multifocal CIDP, with uniform and proximal nerve thickening in the former and segmental nerve enlargement at non-entrapment sites in the latter. These findings are consistent with the literature and support the pathophysiology of segmental inflammation in both diseases [8,9]. The sensitivity of the US in hereditary polyneuropathy was lower compared to the literature [9,10]. This can be explained by the inclusion of other types of CMT, not only CMT 1A. The frequent diagnosis in this cohort was also hereditary neuropathy with liability to pressure palsies (HNPP). In that condition, nerve enlargement is less pronounced and frequent at the entrapment sites [7]. The sensitivity in diabetic polyneuropathy was similar to that reported in the literature. In this condition, nerve US generally shows normal or slightly enlarged nerves, particularly the proximal median and tibial nerve segments [10]. Sensitivity was lowest in idiopathic polyneuropathy, where focal nerve thickening at entrapment and non-entrapment sites was the most common finding.

The US provided a novel diagnosis in 7 (3%) of our patients, including 2 patients with unexpected findings of peripheral nerve tumors reported elsewhere [24]. We also found nerve thickening at non-entrapment sites typical of focal demyelinating polyneuropathy in a patient with EDx findings typical of axonal polyneuropathy, a finding reported also by others [25]. This is useful in cases where EDx studies are unable to provide exact anatomical location.

The US was regarded as contributive in 39 (19%) of our patients by providing information on the nerve size and structures outside the nerves. This is particularly important in patients with polyneuropathy and superimposed focal neuropathy (e.g., patients with diabetic polyneuropathy and suspected median entrapment neuropathy at the wrist). Focal nerve thickening was also found in several patients with amyloid polyneuropathy [20], other metabolic (uremic, vitamin B12 deficiency), autoimmune (rheumatoid arthritis) polyneuropathies, and polyneuropathy of unclear etiology. In MMN and multifocal CIDP arm nerves were characteristically thickened at non-entrapment sites, often without concomitant conduction block on EDx, as reported before [26]. On the other hand, we found nerve thinning in 3 patients with multifocal inflammatory neuropathies (MMN and multifocal CIDP), probably due to secondary axonal loss in the advanced disease stage [18].

The US was confirmative in 50 (25%) of our patients, which increased our confidence in clinical and EDx diagnosis. The high sensitivity of US in CIDP with its well-recognized non-uniform generalized nerve thickening is not unexpected [9]. Together with proximal, particularly tibial and median nerve thickening found in subacute diabetic polyneuropathies [10], this helped us to diagnose superimposed CIDP in patients with diabetes, which is a common and clinically relevant diagnostic dilemma [19,27]. The proximal median nerve segment is commonly swollen also in several other polyneuropathies, including vasculitic, and transthyretin (TTR) amyloid polyneuropathy [20].

In the remaining 52% of patients US did not provide relevant additional information. This occurred mainly in patients with idiopathic polyneuropathy, probable hereditary axonal polyneuropathies (CMT2), and in a minority of patients with diabetic polyneuropathy. These patients had predominantly normal US studies or incidental abnormalities, as reported previously. The majority of acquired and hereditary axonal polyneuropathies have normal or only slightly enlarged nerves [28].

The high frequency of diabetic polyneuropathy in our study cohort is not surprising, as diabetes is the most common cause of polyneuropathy worldwide [29]. By contrast, in our study cohort, hereditary polyneuropathies were probably overrepresented, mainly due to the well-known ability of the US to demonstrate uniform nerve thickening in patients with demyelinating forms [9].

Referral for additional focal neuropathy in patients with generalized polyneuropathy is clinically very useful due to US high sensitivity. By contrast, EDx has higher specificity. Differentiation between inherited and acquired demyelinating polyneuropathies is also useful, and US has high sensitivity and specificity, allowing the exclusion of hereditary demyelinating polyneuropathies. Patients with multifocal CIDP showed the highest US sensitivity and specificity and thus benefited the most. They typically had moderate diffuse nerve enlargement with superimposed focal nerve thickening on non-entrapment sites.

The main strength of this study is the evaluation of the relationship between EDx and US findings in the context of differential diagnosis. This provides additional insight into the complementary role of US and EDx in the diagnosis of polyneuropathy. Calculation of the sensitivity confirmed this important complementary role, especially in hereditary and acquired demyelinating polyneuropathies. In clinical practice, the increased sensitivity allows us to detect not only EDx silent conduction blocks [24] but also segmental nerve thickening in asymptomatic nerves [30].

The present study has several limitations. The most important is its retrospective design, which precluded the standardization of the obtained information. As a consequence, we were not able to provide an objective quantitative assessment of the contribution of the US to patient management. Furthermore, investigators performing US were not blinded to the findings of the neurologic examination and EDx testing. This might introduced certain bias in the evaluation, but on the other side modelled the usual clinical evaluation. US imaging is also dependent on the operator's skill and experience. However, in the present study, we relied mainly on robust parameters of nerve CSAs determined on standard measuring points. Therefore, we do not believe that the operator's skill and experience presented a major issue in the evaluation of major limb nerves. CSA may change throughout the course of the disease, but the study design did not enable us to explore CSA changes over time. Nevertheless, as in clinical practice, we performed US studies at the time of patient referral.

5. Conclusions

This retrospective analysis confirmed the utility of the US in the diagnostic evaluation of patients with different forms of polyneuropathy and also in routine clinical practice. Although the US only rarely provided a new diagnosis, in many patients with polyneuropathy US either contributed to or at least confirmed the diagnosis. The US was most useful in patients with well-formulated and meaningful referral questions.

Author Contributions: Conceptualization, S.P.; methodology, S.P.; validation, M.J.; formal analysis, M.J.; data curation, M.J.; writing—original draft preparation, M.J.; writing—review and editing, S.P.; visualization, M.J.; supervision, S.P.; funding acquisition, S.P. All authors have read and agreed to the published version of the manuscript.

Funding: This research was funded by The Slovenian Research and Innovation Agency, grant number P3-0338.

Institutional Review Board Statement: Ethical review and approval were waived for this study due to the retrospective analysis with no direct involvement of patients and data anonymization.

Informed Consent Statement: Patient consent was waived due to the retrospective analysis with no direct involvement of patients.

Data Availability Statement: The authors will make the raw data supporting this article's conclusions available upon request.

Acknowledgments: The authors thank Gregor Omejec, for his help during peripheral nerve ultrasonography.

Conflicts of Interest: The authors declare no conflicts of interest. The funders had no role in the study's design; in the collection, analyses, or interpretation of data; in the writing of the manuscript; or in the decision to publish the results.

Abbreviations

The following abbreviations are used in this manuscript:

ALS	Amyotrophic lateral sclerosis
CIDP	Chronic inflammatory demyelinating polyneuropathy
CMT1A	Charcot-Marie-Tooth disease type 1A
DADS	Distal acquired demyelinating symmetric polyneuropathy
EDx	Electrodiagnosis
MMN	Multifocal motor neuropathy
TTR	Transthyretin
US	Ultrasonography

References

1. Suk, J.I.; Walker, F.O.; Cartwright, M.S. Ultrasonography of peripheral nerves. *Curr. Neurol. Neurosci. Rep.* **2013**, *13*, 328. [CrossRef]
2. Walker, F.; Cartwright, M.S. *Neuromuscular Ultrasound*; Elsevier, Saunders: Philadelphia, PA, USA, 2011; pp. 1–200.
3. Buchberger, W.; Schön, G.; Strasser, K.; Jungwirth, W. High-resolution ultrasonography of the carpal tunnel. *J. Ultrasound Med.* **1991**, *10*, 531–537. [CrossRef]
4. Kerasnoudis, A.; Tsivgoulis, G. Nerve Ultrasound in Peripheral Neuropathies: A Review. *J. Neuroimaging* **2015**, *25*, 528–538. [CrossRef]
5. Lugao, H.B.; Nogueira-Barbosa, M.H.; Marques, W., Jr.; Foss, N.T.; Frade, M.A. Asymmetric Nerve Enlargement: A Characteristic of Leprosy Neuropathy Demonstrated by Ultrasonography. *PLoS Negl. Trop. Dis.* **2015**, *9*, e0004276. [CrossRef]
6. Zaidman, C.M.; Al-Lozi, M.; Pestronk, A. Peripheral nerve size in normals and patients with polyneuropathy: An ultrasound study. *Muscle Nerve* **2009**, *40*, 960–966. [CrossRef]
7. Zaidman, C.M.; Harms, M.B.; Pestronk, A. Ultrasound of inherited vs. acquired demyelinating polyneuropathies. *J. Neurol.* **2013**, *260*, 3115–3121. [CrossRef]
8. Beekman, R.; van den Berg, L.H.; Franssen, H.; Visser, L.H.; van Asseldonk, J.T.; Wokke, J.H. Ultrasonography shows extensive nerve enlargements in multifocal motor neuropathy. *Neurology* **2005**, *65*, 305–307. [CrossRef]
9. Grimm, A.; Vittore, D.; Schubert, V.; Lipski, C.; Heiling, B.; Decard, B.F.; Axer, H. Ultrasound pattern sum score, homogeneity score and regional nerve enlargement index for differentiation of demyelinating inflammatory and hereditary neuropathies. *Clin. Neurophysiol.* **2016**, *127*, 2618–2624. [CrossRef]
10. Telleman, J.A.; Herraets, I.J.; Goedee, H.S.; van Asseldonk, J.T.; Visser, L.H. Ultrasound scanning in the diagnosis of peripheral neuropathies. *Pract. Neurol.* **2021**, *21*, 186–195. [CrossRef]
11. Grimm, A.; Décard, B.F.; Schramm, A.; Pröbstel, A.K.; Rasenack, M.; Axer, H.; Fuhr, P. Ultrasound and electrophysiologic findings in patients with Guillain-Barré syndrome at disease onset and over a period of six months. *Clin. Neurophysiol.* **2016**, *127*, 1657–1663. [CrossRef]
12. Grimm, A.; Décard, B.F.; Athanasopoulou, I.; Schweikert, K.; Sinnreich, M.; Axer, H. Nerve ultrasound for differentiation between amyotrophic lateral sclerosis and multifocal motor neuropathy. *J. Neurol.* **2015**, *262*, 870–880. [CrossRef] [PubMed]

13. Kerasnoudis, A.; Pitarokoili, K.; Haghikia, A.; Gold, R.; Yoon, M.S. Nerve ultrasound protocol in differentiating chronic immune-mediated neuropathies. *Muscle Nerve* **2016**, *54*, 864–871. [CrossRef] [PubMed]
14. Grimm, A.; Decard, B.F.; Axer, H.; Fuhr, P. The Ultrasound pattern sum score—UPSS. A new method to differentiate acute and subacute neuropathies using ultrasound of the peripheral nerves. *Clin. Neurophysiol.* **2015**, *126*, 2216–2225. [CrossRef]
15. Haffey, P.R.; Bansal, N.; Kaye, E.; Ottestad, E.; Aiyer, R.; Noori, S.; Gulati, A. The Regenerative Potential of Therapeutic Ultrasound on Neural Tissue: A Pragmatic Review. *Pain Med.* **2020**, *21*, 1494–1506. [CrossRef]
16. di Biase, L.; Falato, E.; Caminiti, M.L.; Pecoraro, P.M.; Narducci, F.; Di Lazzaro, V. Focused Ultrasound (FUS) for Chronic Pain Management: Approved and Potential Applications. *Neurol. Res. Int.* **2021**, *2021*, 8438498. [CrossRef]
17. Phenix, C.P.; Togtema, M.; Pichardo, S.; Zehbe, I.; Curiel, L. High intensity focused ultrasound technology, its scope and applications in therapy and drug delivery. *J. Pharm. Pharm. Sci.* **2014**, *17*, 136–153. [CrossRef] [PubMed]
18. Padua, L.; Aprile, I.; Pazzaglia, C.; Frasca, G.; Caliandro, P.; Tonali, P.; Martinoli, C. Contribution of ultrasound in a neurophysiological lab in diagnosing nerve impairment: A one-year systematic assessment. *Clin. Neurophysiol.* **2007**, *118*, 1410–1416. [CrossRef]
19. Van den Bergh, P.Y.K.; van Doorn, P.A. European Academy of Neurology/Peripheral Nerve Society guideline on diagnosis and treatment of chronic inflammatory demyelinating polyradiculoneuropathy: Report of a joint Task Force-Second revision. *Eur. J. Neurol.* **2021**, *28*, 3556–3583. [CrossRef]
20. Podnar, S.; Sarafov, S.; Tournev, I.; Omejec, G.; Zidar, J. Peripheral nerve ultrasonography in patients with transthyretin amyloidosis. *Clin. Neurophysiol.* **2017**, *128*, 505–511. [CrossRef]
21. Padua, L.; Liotta, G.; Di Pasquale, A.; Granata, G.; Pazzaglia, C.; Caliandro, P.; Martinoli, C. Contribution of ultrasound in the assessment of nerve diseases. *Eur. J. Neurol.* **2012**, *19*, 47–54. [CrossRef]
22. Omejec, G.; Podnar, S. Utility of nerve conduction studies and ultrasonography in ulnar neuropathies at the elbow of different severity. *Clin. Neurophysiol.* **2020**, *131*, 1672–1677. [CrossRef]
23. Omejec, G.; Žgur, T.; Podnar, S. Diagnostic accuracy of ultrasonographic and nerve conduction studies in ulnar neuropathy at the elbow. *Clin. Neurophysiol.* **2015**, *126*, 1797–1804. [CrossRef]
24. Podnar, S. Ultrasonography of peripheral nerve tumours: A case series. *Radiol. Oncol.* **2023**, *57*, 35–41. [CrossRef]
25. Herraets, I.J.T.; Goedee, H.S.; Telleman, J.A.; van Eijk, R.P.A.; van Asseldonk, J.T.; Visser, L.H.; van den Berg, L.H.; van der Pol, W.L. Nerve ultrasound improves detection of treatment-responsive chronic inflammatory neuropathies. *Neurology* **2020**, *94*, e1470–e1479. [CrossRef]
26. Li, Y.; Niu, J.; Liu, T.; Ding, Q.; Wu, S.; Guan, Y.; Cui, L.; Liu, M. Conduction Block and Nerve Cross-Sectional Area in Multifocal Motor Neuropathy. *Front. Neurol.* **2019**, *10*, 1055. [CrossRef]
27. Rajabally, Y.A.; Stettner, M.; Kieseier, B.C.; Hartung, H.P.; Malik, R.A. CIDP and other inflammatory neuropathies in diabetes—Diagnosis and management. *Nat. Rev. Neurol.* **2017**, *13*, 599–611. [CrossRef]
28. Breiner, A.; Ebadi, H.; Bril, V.; Barnett, C.; Katzberg, H.D. Ultrasound in Multifocal Motor Neuropathy: Clinical and Electrophysiological Correlations. *J. Clin. Neuromuscul. Dis.* **2019**, *20*, 165–172. [CrossRef]
29. Feldman, E.L.; Callaghan, B.C.; Pop-Busui, R.; Zochodne, D.W.; Wright, D.E.; Bennett, D.L.; Bril, V.; Russell, J.W.; Viswanathan, V. Diabetic neuropathy. *Nat. Rev. Dis. Primers* **2019**, *5*, 42. [CrossRef] [PubMed]
30. Podnar, S. Ultrasonographic abnormalities in clinically unaffected nerves of patients with non-vasculitic motor and sensorimotor mononeuropathies. *Muscle Nerve* **2024**, *70*, 766–773. [PubMed]

applied sciences

MDPI

Article

Morphological Analysis of US Treated PANC-1 Spheroids

Martina Ricci *,†, Mattia Dimitri †, Martina Serio and Andrea Corvi

Department of Industrial Engineering (DIEF), University of Florence, Via di Santa Marta 3, 50139 Florence, Italy; mattia.dimitri@unifi.it (M.D.); martina.serio@edu.unifi.it (M.S.); andrea.corvi@unifi.it (A.C.)
* Correspondence: martina.ricci1@unifi.it; Tel.: +39-3311730200
† These authors contributed equally as first authors.

Abstract: This study investigates the impact of low-intensity continuous ultrasound (LICU) on pancreatic adenocarcinoma (PANC-1) spheroids, emphasizing morphological and volumetric transformations. PANC-1 spheroids were cultured and treated with LICU across frequencies from 1 to 5 MHz. Cell viability and mortality were analyzed through Calcein AM/PI staining, while volumetric and morphological changes were quantified across frequencies from 2 to 4 MHz using advanced imaging techniques and computational tools, including a custom Python OpenCv Library, AnaSP 3.0, a MATLAB based open source tool. Notably, a frequency of 3.5 MHz yielded optimal outcomes, also achieving a reduction in spheroid volume and mortality while minimizing disgregation, a factor linked to metastasis risk. These findings underscore LICU's potential as an effective therapeutic strategy, balancing tumor reduction with the preservation of structural cohesion. The study establishes a methodological framework for optimizing LICU parameters, presenting a less invasive avenue for improving therapeutic outcomes in pancreatic cancer treatment.

Keywords: LICU; volume evaluation; US cancer therapy; computer vision; spheroids disgregation; frequency induced mortality

1. Introduction

Low-intensity continuous ultrasound (LICU) is gaining attention as a promising therapeutic strategy in oncology, particularly for treating challenging malignancies such as pancreatic ductal adenocarcinoma (PDAC). Unlike high-intensity focused ultrasound (HIFU), which relies on thermal ablation, LICU operates through non-thermal mechanisms, offering a fundamentally different approach to cancer treatment [1,2].

HIFU achieves tissue destruction by delivering high-intensity and high-voltage ultrasound waves to generate localized heating, resulting in thermal damage that ablates both cancerous and, potentially, adjacent healthy tissues [3–5]. In contrast, LICU uses significantly lower intensities that avoid thermal effects. Instead, it leverages pressure-related phenomena such as acoustic streaming and cavitation. Acoustic streaming refers to fluid motion induced by ultrasound waves, while cavitation involves the formation and oscillation of microbubbles within tissues [6,7].

Additionally, LICU has been associated with frequency-based resonance phenomena, a topic explored in recent studies [8]. Furthermore, LICU can also be associated with inhibition of tumor cell proliferation and enhancing anticancer drugs delivery through a phenomenon called sonoporation that enhances drug delivery through cell membranes [9–14].

Preliminary research [2,14–16] suggests that it can induce cytotoxic effects in cancer cells by disrupting their internal structures and communication networks. However,

https://doi.org/10.3390/app15041707

its precise mechanisms of action remain under investigation, particularly regarding the interplay between pressure, frequency, and cellular responses.

This study aims to establish the optimal frequency parameters for LICU stimulation and to explore the correlation between spheroid mortality and the volumetric as well as topological changes observed under specific experimental conditions.
The rationale behind this approach is to determine whether frequency-based mortality in LICU can be effective and correlated with a reduction in spheroid volume, providing additional insights into the mechanisms influencing spheroid dynamics.

Additionally, this analysis aims to prevent scenarios where, despite increased mortality, the spheroid mass expands or diffuses, potentially undermining therapeutic effectiveness. The US testing setup in this experiment was calibrated to deliver ultrasound to 3D cancer spheroids through a water medium. Mortality data were obtained using Calcein AM/PI staining [17–20], while morphological and volumetric changes were analyzed using computer vision tools integrated with advanced imaging techniques.

Although previous research has suggested a link between spheroid volume and mortality [21–23], the interplay between these factors under ultrasound stimulation remains insufficiently understood. This study specifically investigates the correlation between low-intensity continuous ultrasound stimulation (under 3, W/cm^2) and the evolution of spheroid morphology and volume. Confocal microscopy images were analyzed using computer vision-based morphometric techniques to quantify these changes post-stimulation.

The use of three-dimensional spheroid models has become essential in cancer biology research, as these models replicate key characteristics of solid tumors, including gradients of oxygen, nutrients, and therapeutic agents when incorporated into the study. This work aims to establish a relationship between spheroid mortality and volumetric changes under ultrasound stimulation, focusing on parameters such as volume, jaggedness, and compactness.

Additionally, this study aims to correlate the effects of ultrasound treatment intensity with those observed in a previous protocol [1]. The ultrasound stimulation employed here falls within the category of non-thermal therapeutic ultrasound, consistent with recent advancements in the field. As highlighted in the comprehensive study by O'Reilly et al. on non-thermal therapeutic ultrasound [24], the treatment parameters of this study are well-positioned within the low-intensity range, including pressure levels.

The pressure, voltage and frequency tested in [1] are visually reassumed in Figure 1. In contrast, this study investigates the effects of LICU at greater range of frequencies, enabling a direct comparison between the two protocols.

In the experiments conducted in the study by Dimitri et al. [1], the conclusion was that further investigation was required to determine whether there was potential expansion in the volume of the speroids following stimulation. This work therefore sought not only to test a wider range of frequencies, but also to assess the occurrence of volume expansion.

Commonly employed methods for the analysis of spheroids and their features encompass a variety of techniques. These methods include:

- Brightfield microscopy: Enables high-resolution imaging of spheroid morphology and facilitates real-time observation during the growth phase.
- Confocal microscopy: Allows for three-dimensional visualization through z-stack imaging and enables fluorescence staining, making it possible to visualize molecules or detect differences in cell structures with high specificity.
- Feature assessment technique: These often rely on image segmentation and 3D reconstruction, typically implemented using software tools such as ImageJ, MATLAB, or other dedicated applications. Feature extraction can also be performed using programming languages like Python, leveraging advanced libraries such as OpenCV [25–29].

- Emerging technologies: Optical coherence tomography (OCT) is becoming increasingly popular for its ability to provide non-invasive, high-resolution volumetric measurements [30,31]. These methods are often complemented by quantitative metrics such as spheroid diameter, volume, surface roughness, and compactness, which are essential for understanding the biological and mechanical properties of these structures. Regarding this mechanical property of the pancreas, it would be interesting to investigate how much the healthy and malignant properties of the tissue differ between its native form and the spheroid form. This topic has been studied only in tissue form [32], but some studies have recently been published on single cells [33]. However, there is still a lack of information regarding spheroids, partly because their analysis requires a humid and liquid suspension environment. Additionally, working with Atomic Force Microscopy (AFM) under such conditions is not trivial, as it introduces technical challenges related to stability, resolution, and sample preparation

Together, these methodologies offer a comprehensive toolkit for characterizing spheroids, providing valuable insights into their growth dynamics, structural changes, and responses to experimental treatments.

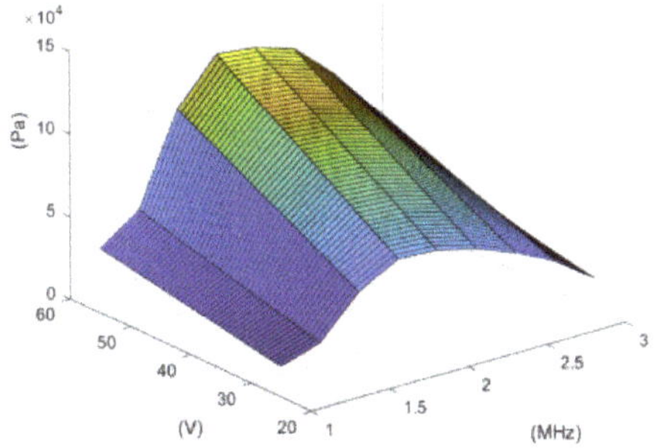

Figure 1. This image shows a 3D representation of pressure, voltage, and frequency levels during the previous stimulation protocol [1], generated using MATLAB from experimental data obtained by HGL200 hydrophone from Honda instruments. This visualization provides an overview of the treatment parameters before the updated protocols. The color changes refer to the peak in the variables intensity.

2. Materials and Methods

2.1. Spheroid Culturing and Staining

Spheroids were cultured using the hanging drop method, a widely recognized technique for three-dimensional cell culture. PANC-1 cells were seeded at a density of 1000 cells per drop in Dulbecco's Modified Eagle Medium (DMEM) supplemented with 10% fetal bovine serum (FBS) and 1% penicillin-streptomycin. The hanging drops were incubated at 37 °C in a 5% CO_2 atmosphere for 48 h to facilitate spheroid formation. Cell viability and mortality within the spheroids were assessed using Calcein AM and Propidium Iodide (PI) staining. Calcein AM labeled live cells with green fluorescence, while PI marked dead cells with red fluorescence. Following minor modifications to the manufacturer's protocol, spheroids were incubated in a staining solution containing 2 µM Calcein AM and 4 µM PI for 30 min at 37 °C before imaging. The total number of samples analyzed, including both treated spheroids and controls, was 256.

2.2. Microscopy and Imaging

Depth-resolved imaging of spheroids was performed using a Zeiss LSM 880 confocal microscope (Carl Zeiss AG, Oberkochen, Germany) equipped with a 20× objective. Z-stack imaging captured sections with a thickness of 5 μm, with specific channels for Calcein AM (488 nm excitation, 515 nm emission) and PI (561 nm excitation, 617 nm emission). All imaging parameters, including laser power, gain, and exposure time, were kept constant across samples. Calibration beads were employed to verify imaging conditions.

2.3. Ultrasound Stimulation

Spheroids were exposed to low-intensity continuous ultrasound (LICU) stimulation using a function generator (Hantek DSO4102C, Qingdao, China) connected to the trasducer immerged in a tank of degassed water manteined at 34°C to avoid thermal shock in the tridimensional coltures. Two different transducers were used to cover a frequency range from 1 MHz to 5 MHz. The US probes used in the protocol were selected for their central frequency bands: the A307S-SU immersion transducer with a 5 MHz central frequency (Evident Scientific, Waltham, MA, USA) and the WS75-2 transducer (Ultran Group, State College, PA, US). Both transducers are non-focalized cylindrical immersion probes. Uniform energy delivery across samples was ensured using a mechanical two-axis positioning system controlled by Arduino.

2.4. Image Segmentation and Mortality Analysis

Image segmentation and subsequent mortality analysis were carried out using a custom Python-based algorithm, leveraging OpenCV functions [34]. This approach, designed to optimize the segmentation of spheroid confocal images and extract meaningful biological parameters, was implemented through several sequential steps commonly applied in medical image analysis [35–37].

- Image Preprocessing: The original RGB images were converted to grayscale using OpenCV's `cvtColor` function, a standard preprocessing preparatory step that reduces computational complexity while retaining essential information of the image [38,39]. Gaussian blurring was then applied to suppress noise and enhance the precision of subsequent steps. This denoising technique is critical in many image processing applications, particularly in biological imaging, where noise reduction is essential for accurate segmentation [40,41].
- Thresholding: A thresholding procedure was applied to the grayscale image using a method already present in the OpenCV library. This process is a common technique for isolating objects from the background, which is essential in imaging tasks such as cell segmentation [36,42,43]. Pixels below the average grayscale level were set to white, effectively isolating the spheroid and creating a binary representation for further analysis.
- Edge Detection and Contour Identification: The edges of the binary mask were first detected using the Canny edge detection algorithm [44,45]. This algorithm identifies areas of the image where there is a sharp contrast in pixel intensity, essentially outlining the boundaries of objects. Following this, the OpenCV findContours function was used to detect the shapes or contours formed by these edges. Among the detected contours, the largest one, assumed to correspond to the spheroid, was selected and stored for further analysis.
- Mask binary generation and removal of the background of the original confocal color image: A binary mask was obtained by overlaying the detected contour onto a black background using the OpenCV drawContours function. This resulted in an image where the contour was white (255) and the background black (0). The binary mask was then applied to the original color image; this operation performed a pixel-wise logical AND between the binary mask and the color image, effectively isolating and displaying only the region corresponding to the spheroid, while all other areas were set to black.

3. Mortality Analysis

The segmented images with the removal of the background as in Figure 2 were analyzed to assess cell viability. The area of the spheroid was determined by counting the white pixels in the binary mask using the countNonZero function, which provides the total number of pixels within the spheroid region.

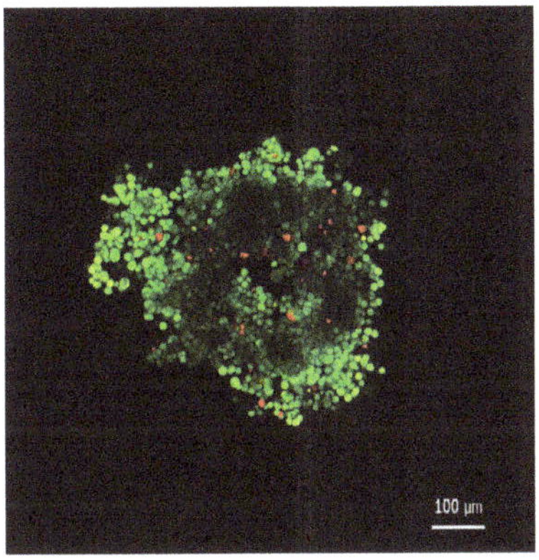

Figure 2. This image obtained by confocal microscopy shows how the Calecin/PI stained image with the background removal appears. The red color is due to PI staining while the green color is due to calcein staining, indicating the mortality and the viability of the single cells.

Subsequently, the ratio of red pixels (representing Propidium Iodide-stained cells) to the total pixels within the spheroid area was calculated. This ratio, indicative of apoptotic cells, was used to quantify the effects of ultrasound treatment, providing insights into cell mortality rates within the treated spheroids. The normalization to the spheroid area was performed to ensure the results were not absolute but relative to the size of the samples, accounting for the natural variability in dimensions observed in biological specimens.

4. Depth-Resolved Mortality Analysis

To study the variation of cell mortality along the spheroid's depth, confocal microscopy Z-stack images were processed. The Figure 3 shows how the spheroid appears on the three sections from the Z-stack. The algorithm was adapted to analyze each image slice individually, enabling the reconstruction of mortality profiles as a function of depth. Furthermore, the advanced version of the segmentation script automatically indexed the slices, streamlining the integration of depth-specific data into the overall statistical analysis.

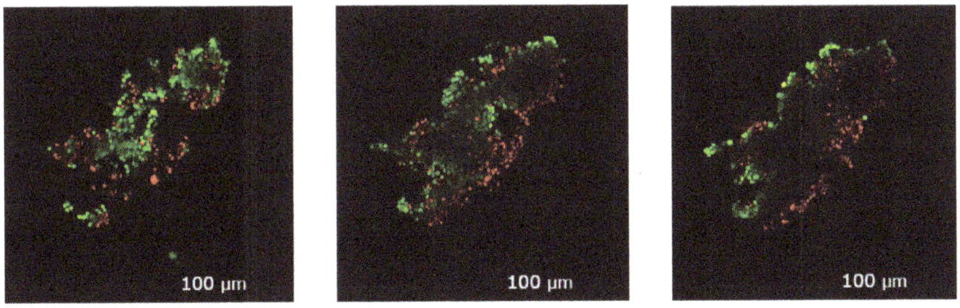

Figure 3. This figure collects the three section from the Z-stack derivate from the same spheroid. The red color is due to PI staining while the green color i due to calcein staining, indicating the mortality and the viability of the single cells.

5. Volume and Disgregation Index Estimation

The volume of spheroids, measured in voxels, was estimated using the ReViSP software tool [46] integrated in the AnaSP 3.0 open source tool [47]. This tool generates 3D volume reconstructions based on the cross-sectional areas of segmented binary masks derived from Z-stack images processed with custom Python scripts. A graphical representation of the steps described in this section is provided in Figure 4.

The volume of each spheroid was calculated using segmented binary masks. To account for biological variability, the volumes of treated spheroids were subjected to a weighted average relative to the volumes of untreated control spheroids from the same experimental day. This normalization step minimized variability caused by biological differences between samples, ensuring more reliable comparisons. For each treatment frequency, the mean volume of all treated spheroids was computed, allowing for the correlation of volume changes with cell mortality at specific frequencies.

5.1. Morphological Analysis and Disgregation Index

The perimeter-to-area ratio of each spheroid's cross-sectional image was calculated to assess morphological changes. This metric, referred to as the disgregation index, provides a measure of spheroid border irregularity. The disgregation index was evaluated separately for treated and untreated spheroids to determine the extent of cellular dispersion induced by ultrasound (US) excitation. This dispersion may have adverse outcomes, including the potential promotion of metastasis [48–52].

For statistical data analysis, the perimeter-to-area ratios were averaged within each frequency group for both treated and control samples. A ratio of these averages (treated versus control) was calculated to evaluate the impact of treatment on the disgregation index, while accounting for biological variability. Area and perimeter measurements were obtained using AnaSP [53], an open source tool designed for spheroid analysis, with segmented images of the spheroids used as input.

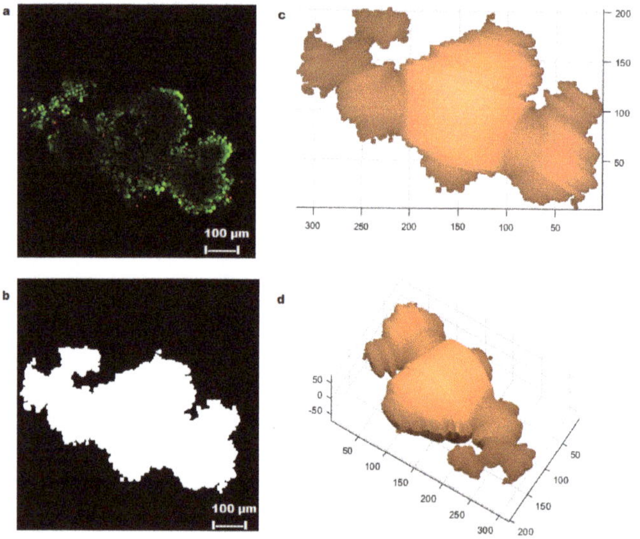

Figure 4. Steps to derive the volume evalutaion: how the varius steps appear graphically (**a**); the confocal image with Calcein/PI staining (**b**); the segmentation and binary mask (**c**); top view from ReVisp graphical interface (**d**), a side view of the volume is visible.

For the disgregation index, statistical analysis was performed by first averaging the ratios for each frequency across both treated and control groups. Then, a ratio of these averages was calculated to evaluate changes in the disgregation index introduced by treatment, while accounting for biological variability.

5.2. Software and Code Availability

AnaSP and ReViSP, two open-source tools, are now integrated into the AnaSP 3.0 open-source framework, which can be accessed via their respective repositories [46,47,53]. Additionally, custom Python scripts developed for segmentation of cofocal images and analysis are available upon request from the corresponding author.

6. Results

The analysis of spheroids subjected to LICU revealed significant correlations between treatment frequency, as shown in Figure 5, and tumor volume reduction, illustrated in Figure 6. Finally, the spheroid disgregation index is displayed in Figure 7. Data were gathered across frequencies of 1 MHz to 5 MHz for the mortality analysis, for the volume and disgregation index we only focused on the range from 2 MHz to 4 MHz. The focus on a narrower range was motivated by the results of the mortality analyses. Specifically, the most significant findings were observed between 2 MHz and 4 MHz, where two distinct peaks in mortality were identified: one at 2 MHz and another, of varying magnitude, at 3.5 MHz.

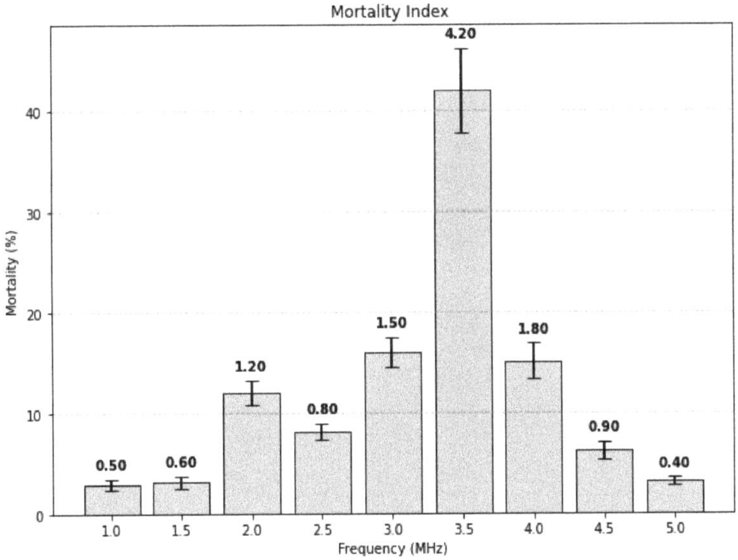

Figure 5. The mortality ratio obtained by evaluating the PI-stained pixels weighted on the spheroid's segmented area. The standard deviation is also displayed for each histogram column.

In the volume histogram, intensity values around 1 indicate that the volume remains unchanged, neither expanding nor contracting. Similarly, in the disaggregation histogram, values close to 1 suggest that the spheroids neither disaggregate nor expand. These values are highlighted in bold in the Figures 6 and 7.

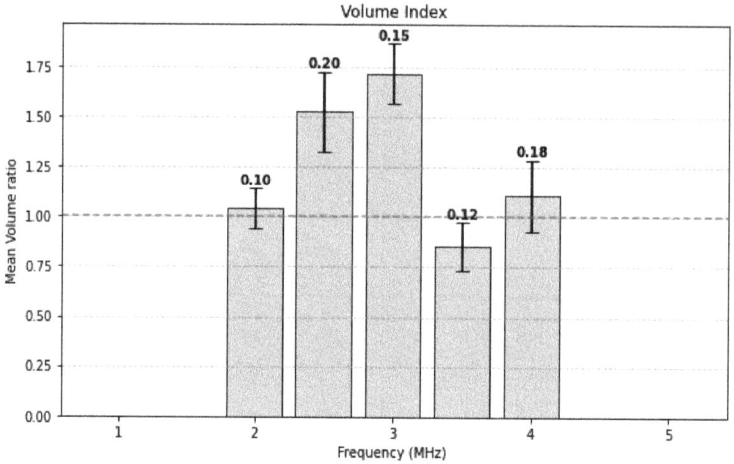

Figure 6. The volume mean ratio obtained by making the evaluation of treated Volume on Control volume. The standard deviation is also displayed for each histogram column.

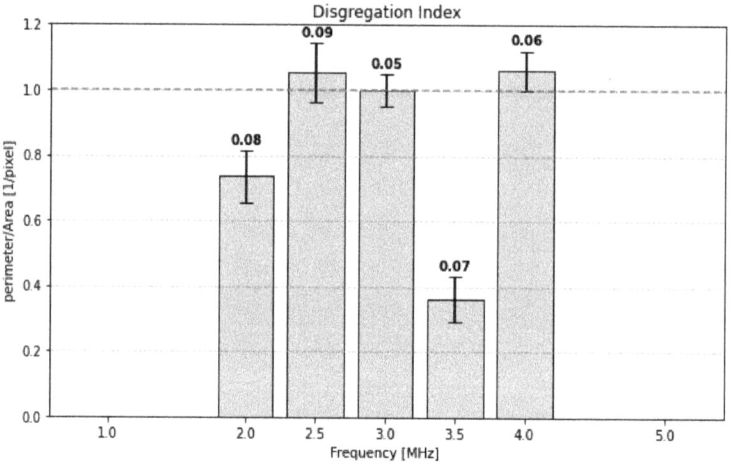

Figure 7. The disgregation index is obtained by making the evaluation of the mean treated perimeter on area on the mean control perimeter on area. The standard deviation is also displayed for each histogram column.

7. Discussion

The results indicate a strong correlation between LICU treatment at specific frequencies and the observed changes in spheroid behavior, as demonstrated in Figure 6.

At a frequency of 2 MHz, a substantial reduction in spheroid volume was recorded. This effect was accompanied by a smaller peak in mortality rates, as depicted in Figure 5, alongside a noticeable decrease in the disgregation index (Figure 7).

Such findings suggest that LICU at 2 MHz induces measurable alterations in three-dimensional spheroid cultures. These observations hint at the potential for off-target effects at 2 MHz, which may influence spheroid architecture without completely compromising cell viability.

In contrast, the effects observed at 3.5 MHz were particularly striking. This frequency was associated with a marked increase in mortality rates, as illustrated in Figure 5, coupled with a marked reduction in the volume of the spheroid (Figure 6).

Furthermore, there is a significant decrease in the disgregation index (Figure 7) at 3.5 MHz underscores. Unlike the effects at 2 MHz, treatment at 3.5 MHz appeared to cause higher mortality rates, still preventing disaggregation.

This distinction is crucial, as excessive disgregation could facilitate metastasis, a concern supported by the literature that highlights the mechanical and structural consequences of spheroid disaggregation [48–51].

Thus, LICU at 3.5 MHz demonstrates a dual capability: it induces significant cellular mortality while promoting spheroid destabilization, with potential implications for therapeutic strategies targeting tumor cohesion and survival.

These findings highlight the nuanced interplay between frequency-specific LICU effects on spheroids. Although 2 MHz elicits subtle structural changes with limited mortality, 3.5 MHz emerges as a more potent frequency to cause cell death, potentially offering a targeted approach to therapeutic interventions. Future investigations should focus on further elucidating the mechanistic underpinnings of these frequency-dependent effects, with an emphasis on optimizing therapeutic windows to balance efficacy and safety.

The following section provides a concise yet comprehensive statistical analysis of the data presented in the figures. This analysis aims to assess the significance of the observed differences and to identify the correlations.

Regarding the mortality data it is noted that the sample size exceeds 250 and the p value is significantly less than 0.05.

Regarding the statistical analysis of the indices, the standard deviation of the volume index is consistently below 0.2, indicating relatively low variability. The statistical correlation values derived from the volume index suggest a significant relationship with mortality.

Specifically, the volume index shows a correlation coefficient of -0.496, indicating a correlation. The corresponding t-score is -3.023 and the p-value for this result is 0.0053, which is well below the commonly used significance threshold of 0.05. This confirms that the correlation is statistically significant and suggests that changes in volume index may be related to mortality.

For the disgregation index, the standard deviation is smaller, remaining below 0.09, suggesting even less variability in these measurements.

This index shows a stronger negative correlation with mortality, with a correlation coefficient of -0.800. The t-score is -7.052. The corresponding p-value is 1.14×10^{-7}, which is extremely small, indicating an exceptionally strong and statistically significant relationship.

The statistical analysis reveals that both the volume index and the disgregation index show significant correlations with mortality. Both indices have p-values well below 0.05, confirming their statistical significance.

These findings suggest that both indices are relevant factors in understanding mortality. Therefore, it indicates that mortality is statistically correlated with the morphological changes detected by the computer vision analysis conducted.

All statistical evaluations were performed using Google Sheets and Python.

At this stage, a comparison with similar studies is essential. The following section provides a review of relevant research.

Bazou et al. [2] published a study in 2017 investigating the effects of LICU treatment. Their findings indicate that continuous low-intensity ultrasound (LICU) increases the levels of pro-inflammatory cytokines IFN-γ, IL-1β, and TNF-α in murine pancreatic tumors without inducing significant thermal, mechanical, or structural alterations. These results suggest that LICU has the potential to modulate the tumor inflammatory response, which could lead to additional anti-tumor effects.

More recently, in 2023, González et al. [14] demonstrated that LICU significantly inhibits the migration of PANC-1 cells in vitro. This finding suggests that LICU could play a role in limiting the spread of cancer cells.

In 2022, Song et al. [54] proposed a mechanistic explanation for the efficacy of low-intensity ultrasound, supported by both in vitro and in vivo experiments. Their study focuses on the mechanosensitive Piezo1 channel as a potential therapeutic target for pancreatic ductal adenocarcinoma (PDAC). Piezo1 is a membrane protein that responds to mechanical stimuli, such as pressure changes, by facilitating calcium (Ca^{2+}) influx into the cell. This calcium signaling regulates key biological processes, including apoptosis, which is critical for cancer therapy.

The results show that Piezo1 is highly expressed in pancreatic cancer cells compared to healthy tissue. The experiments used ultrasound combined with microbubbles (US + MB), a non-invasive technique that uses microbubble oscillations to generate localised mechanical stimuli. This combination activates Piezo1, causing an increase in intracellular calcium and mitochondrial dysfunction, which in turn induces tumour cell apoptosis.

It is important to note that while there is evidence to suggest a correlation between the Piezo1 cation channel and the induction of cell death by LICU, the underlying mechanisms responsible for the observed effectiveness at 3.5 MHz and the secondary peak at 2 MHz remain to be fully elucidated. It is therefore plausible that these phenomena may not be attributable to a single, shared biological rationale.

These studies are closely related to the work presented in this publication, further confirming the results obtained in the experiments. However, certain aspects require further clarification, both in existing literature and in future research involving the use of LICU in the pancreas. Specifically, additional studies on in vivo applications are necessary, along with more comparative tests using healthy cells to validate the selectivity of the treatment.

8. Conclusions

The findings demonstrate that mortality peaks coincide with significant spheroid shrinkage. Moreover, the correlation with the reduced disgregation index highlights the need for further investigation, building on the team's previous work [1], which emphasized the importance of excluding potential disgregation caused by LICU. The simultaneous reduction in volume and mortality at 2 and 3.5 MHz frequencies suggests a therapeutic balance where LICU achieves maximum efficiency in targeting tumor spheroids while minimizing the risks of disgregation. Notably, the decrease in the disgregation index at these critical frequencies further supports LICU's efficacy, indicating an optimal mechanism that compromises tumor cohesion without promoting cellular dispersion. The ultimate goal of this work is to gain a deeper understanding of the effectiveness of LICU. Demonstrating selectivity through in vitro testing on healthy pancreatic cells, along with further validation of its efficacy in murine models inoculated with human PDAC, could significantly enhance the technology readiness level and move this technology closer to feasibility for human treatments. Furthermore, it would be highly valuable to investigate the behavior of the Piezo1 cationic channel during stimulation, focusing on its activation dynamics and the role of calcium ion flux, to provide greater insight into the mechanistic pathways involved and refine the therapeutic strategy. These steps are considered crucial for advancing the research and bringing the technology closer to clinical application.

Despite the promising results obtained, certain limitations must be addressed to further advance the clinical applicability of LICU. One major challenge is the lack of commercially available ultrasonic amplifiers capable of continuously amplifying ultrasound waves for biomedical applications. Currently, the only available solutions are designed for civil and seismic engineering, necessitating the development of a customized system optimized for therapeutic use. Additionally, translating the stinolation protocol to in vivo models presents significant technical hurdles. Water, commonly used as a coupling medium in in vitro experiments, is a difficult fitting option for in vivo applications, necessitating the identification of a biocompatible and echogenic coupling material. Overcoming these limitations will be essential for refining the therapeutic approach and bridging the gap between preclinical validation and potential clinical translation.

Author Contributions: Conceptualization, M.D. and A.C.; methodology, A.C.; software, M.R. and M.S.; validation M.D. and A.C; formal analysis, M.R.; investigation, M.R.; resources, M.D. and A.C.; data curation, M.R.; writing—original draft preparation, M.R.; writing—review and editing, M.R. and A.C.; visualization, M.R.; supervision, M.D. and A.C.; project administration, A.C. and M.D.; funding acquisition, A.C. and M.D. All authors have read and agreed to the published version of the manuscript.

Funding: Financial support from the Italian Ministry of University and Research (PNRR projects "AGE-IT—Ageing well in an ageing society") is gratefully acknowledged.

Institutional Review Board Statement: Ethical review and approval were not required for this study, as confirmed by the Ethical Committee for Animal Experimentation of the University of Florence. The study exclusively involved commercially available cell lines (PANC-1 spheroids) and did not include live animals or patient-derived samples.

Data Availability Statement: The Python code is available upon request from the corresponding author.

Acknowledgments: The authors wish to acknowledge Annarosa Arcangeli and Claudia Duranti, Jessica Iorio, and Chiara Capitani for providing the cell cultures and for their valuable suggestions.

Conflicts of Interest: The authors declare no conflicts of interest.

References

1. Dimitri, M.; Duranti, C.; Aquino, S.; Mazzantini, L.; Iorio, J.; Lulli, M.; Ricci, M.; Capineri, L.; Arcangeli, A.; Corvi, A. Biophysical and Biomechanical Effect of Low Intensity US Treatments on Pancreatic Adenocarcinoma 3D Cultures. *Appl. Sci.* **2022**, *12*, 666. [CrossRef]
2. Bazou, D.; Maimon, N.; Munn, L.L.; Gonzalez, I. Effects of low intensity continuous ultrasound (LICU) on mouse pancreatic tumor explants. *Appl. Sci.* **2017**, *7*, 1275. [CrossRef]
3. Maloney, E.; Hwang, J.H. Emerging HIFU applications in cancer therapy. *Int. J. Hyperth.* **2015**, *31*, 302–309. [CrossRef] [PubMed]
4. Barkin, J. High intensity focused ultrasound (HIFU). *Can. J. Urol.* **2011**, *18*, 5634.
5. Haar, G.T. HIFU Tissue Ablation: Concept and Devices. In *Therapeutic Ultrasound*; Escoffre, J.M., Bouakaz, A., Eds.; Springer: Cham, Switzerland, 2016; pp. 3–20.
6. Feril, L.B.; Kondo, T. Biological effects of low intensity ultrasound: The mechanism involved, and its implications on therapy and on biosafety of ultrasound. *J. Radiat. Res.* **2004**, *45*, 479–489. [CrossRef]
7. Wood, A.K.; Sehgal, C.M. A review of low-intensity ultrasound for cancer therapy. *Ultrasound Med. Biol.* **2015**, *41*, 905–928. [CrossRef] [PubMed]
8. Fraldi, M.; Cugno, A.; Deseri, L.; Dayal, K.; Pugno, N. A frequency-based hypothesis for mechanically targeting and selectively attacking cancer cells. *J. R. Soc. Interface* **2015**, *12*, 20150656. [CrossRef]
9. Loria, R.; Giliberti, C.; Bedini, A.; Palomba, R.; Caracciolo, G.; Ceci, P.; Falvo, E.; Marconi, R.; Falcioni, R.; Bossi, G.; et al. Very low intensity ultrasounds as a new strategy to improve selective delivery of nanoparticles-complexes in cancer cells. *J. Exp. Clin. Cancer Res.* **2019**, *38*, 1–11. [CrossRef] [PubMed]
10. Ricci, M.; Barbi, E.; Dimitri, M.; Duranti, C.; Arcangeli, A.; Corvi, A. Sonoporation, a Novel Frontier for Cancer Treatment: A Review of the Literature. *Appl. Sci.* **2024**, *14*, 515. [CrossRef]
11. Zhang, K.; Xu, H.; Chen, H.; Jia, X.; Zheng, S.; Cai, X.; Wang, R.; Mou, J.; Zheng, Y.; Shi, J. CO_2 bubbling-based'nanobomb'system for targetedly suppressing panc-1 pancreatic tumor via low intensity ultrasound-activated inertial cavitation. *Theranostics* **2015**, *5*, 1291. [CrossRef]

12. Kuo, Y.Y.; Chen, W.T.; Lin, G.B.; Lu, C.H.; Chao, C.Y. Study on the effect of a triple cancer treatment of propolis, thermal cycling-hyperthermia, and low-intensity ultrasound on PANC-1 cells. *Aging* **2023**, *15*, 7496. [CrossRef] [PubMed]
13. Qiu, F.; Chen, J.; Cao, J.; Diao, F.; Huang, P. Low-intensity low-frequency ultrasound enhances the chemosensitivity of gemcitabine-resistant ASPC-1 cells via PI3K/AKT/NF-κB pathway-mediated ABC transporters. *Oncol. Rep.* **2020**, *44*, 1158–1168. [CrossRef]
14. González, I.; Luzuriaga, J.; Valdivieso, A.; Candil, M.; Frutos, J.; López, J.; Hernández, L.; Rodríguez-Lorenzo, L.; Yagüe, V.; Blanco, J.L.; et al. Low-intensity continuous ultrasound to inhibit cancer cell migration. *Front. Cell Dev. Biol.* **2023**, *10*, 842965. [CrossRef] [PubMed]
15. Lagneaux, L.; de Meulenaer, E.C.; Delforge, A.; Dejeneffe, M.; Massy, M.; Moerman, C.; Hannecart, B.; Canivet, Y.; Lepeltier, M.F.; Bron, D. Ultrasonic low-energy treatment: A novel approach to induce apoptosis in human leukemic cells. *Exp. Hematol.* **2002**, *30*, 1293–1301. [CrossRef]
16. Amaya, C.; Luo, S.; Baigorri, J.; Baucells, R.; Smith, E.R.; Xu, X.X. Exposure to low intensity ultrasound removes paclitaxel cytotoxicity in breast and ovarian cancer cells. *BMC Cancer* **2021**, *21*, 1–13. [CrossRef] [PubMed]
17. Dong, G.; Wang, S.; Ge, Y.; Deng, Q.; Cao, Q.; Wang, Q.; Shang, Z.; OuYang, W.; Li, J.; Liu, C.; et al. Serum-free culture system for spontaneous human mesenchymal stem cell spheroid formation. *Stem Cells Int.* **2019**, *2019*, 6041816. [CrossRef]
18. Kiyomi, A.; Miyakawa, R.; Matsumoto, J.; Yamazaki, K.; Imai, S.; Yuan, B.; Hirano, T.; Sugiura, M. Potent antitumor activity of cepharanthine against triple-negative breast cancer spheroids compared with tetrandrine. *Oncol. Lett.* **2020**, *20*, 331. [CrossRef] [PubMed]
19. Singh, S.K.; Abbas, S.; Saxena, A.K.; Tiwari, S.; Sharma, L.K.; Tiwari, M. Critical role of three-dimensional tumorsphere size on experimental outcome. *Biotechniques* **2020**, *69*, 333–338. [CrossRef] [PubMed]
20. Kessel, S.; Cribbes, S.; Déry, O.; Kuksin, D.; Sincoff, E.; Qiu, J.; Chan, L.L.Y. High-throughput 3D tumor spheroid screening method for cancer drug discovery using celigo image cytometry. *Slas Technol. Transl. Life Sci. Innov.* **2017**, *22*, 454–465. [CrossRef]
21. Yeon, S.E.; No, D.Y.; Lee, S.H.; Nam, S.W.; Oh, I.H.; Lee, J.; Kuh, H.J. Application of concave microwells to pancreatic tumor spheroids enabling anticancer drug evaluation in a clinically relevant drug resistance model. *PLoS ONE* **2013**, *8*, e73345. [CrossRef]
22. Dinter, J.; Friedrich, R.P.; Yang, H.; Pilarsky, C.; Mangge, H.; Pöttler, M.; Janko, C.; Alexiou, C.; Lyer, S. Mitoxantrone and mitoxantrone-loaded iron oxide nanoparticles induce cell death in human pancreatic ductal adenocarcinoma cell spheroids. *Materials* **2023**, *16*, 2906. [CrossRef] [PubMed]
23. Maftouh, M.; Avan, A.; Sciarrillo, R.; Granchi, C.; Leon, L.G.; Rani, R.; Funel, N.; Smid, K.; Honeywell, R.; Boggi, U.; et al. Synergistic interaction of novel lactate dehydrogenase inhibitors with gemcitabine against pancreatic cancer cells in hypoxia. *Br. J. Cancer* **2014**, *110*, 172–182. [CrossRef] [PubMed]
24. O'Reilly, M.A. Exploiting the mechanical effects of ultrasound for noninvasive therapy. *Science* **2024**, *385*, eadp7206. [CrossRef] [PubMed]
25. Li, G.; Zhang, Y.; Xu, B.; Li, X. Image analysis and processing of skin cell injury based on OpenCV. *J. Phys. Conf. Ser.* **2019**, *1237*, 032003. [CrossRef]
26. Meimban, R.J.; Fernando, A.R.; Monsura, A.; Rañada, J.; Apduhan, J. Blood cells counting using python OpenCV. In Proceedings of the 2018 14th IEEE International Conference on Signal Processing (ICSP), Piscataway, NJ, USA, 12–16 August 2018; Volume 144, pp. 50–53.
27. Khan, A.A. Tuberculosis: Image Segmentation Approach Using OpenCV. *Sukkur IBA J. Comput. Math. Sci.* **2018**, *2*, 1–7.
28. Ojeme, B.; Quinn, F.; Karls, R.; Quinn, S. Fully Automated Methods for the Detection and Segmentation of Mitochondria in Microscopy Images. *Int. J. Mech. Mater. Eng.* **2023**, *17*, 42–47.
29. Bui, M.Q.V.; Ngo, D.T.; Pham, H.A.; Nguyen, D.D. GAC3D: Improving monocular 3D object detection with ground-guide model and adaptive convolution. *PeerJ Comput. Sci.* **2021**, *7*, e686. [CrossRef]
30. Huang, Y.; Zou, J.; Badar, M.; Liu, J.; Shi, W.; Wang, S.; Guo, Q.; Wang, X.; Kessel, S.; Chan, L.L.Y.; et al. Longitudinal morphological and physiological monitoring of three-dimensional tumor spheroids using optical coherence tomography. *J. Vis. Exp. JoVE* **2019**, *144*, e59020. [CrossRef]
31. Sharma, M.; Verma, Y.; Rao, K.; Nair, R.; Gupta, P. Imaging growth dynamics of tumour spheroids using optical coherence tomography. *Biotechnol. Lett.* **2007**, *29*, 273–278. [CrossRef] [PubMed]
32. Pagnanelli, M.; De Gaetano, F.; Callera, A.; Nappo, G.; Capretti, G.; Carrara, S.; Ferrari, A.M.R.; Cellesi, F.; Costantino, M.L.; Zerbi, A. Analysis of the Mechanical Characteristics of Human Pancreas through Indentation: Preliminary In Vitro Results on Surgical Samples. *Biomedicines* **2024**, *12*, 91. [CrossRef] [PubMed]
33. Nguyen, A.V.; Nyberg, K.D.; Scott, M.B.; Welsh, A.M.; Nguyen, A.H.; Wu, N.; Hohlbauch, S.V.; Geisse, N.A.; Gibb, E.A.; Robertson, A.G.; et al. Stiffness of pancreatic cancer cells is associated with increased invasive potential. *Integr. Biol.* **2016**, *8*, 1232–1245. [CrossRef]
34. Bradski, G.; Kaehler, A. *Learning OpenCV: Computer Vision with the OpenCV Library*; O'Reilly Media, Inc.: Sebastopol, CA, USA, 2008.
35. Haidekker, M. *Advanced Biomedical Image Analysis*; John Wiley & Sons: Hoboken, NJ, USA, 2010.

36. Altun, A.A.; Taghiyev, A. Advanced image processing techniques and applications for biological objects. In Proceedings of the 2017 2nd IEEE International Conference on Computational Intelligence and Applications (ICCIA), Piscataway, NJ, USA, 8–11 September 2017, pp. 340–344.
37. Merchant, F.; Castleman, K. *Microscope Image Processing*; Academic Press: Cambridge, MA, USA, 2022.
38. Sarfraz, M. Introductory Chapter: On Digital Image Processing. In *Digital Imaging*; IntechOpen: London, UK, 2020.
39. Gonzalez, R.C. *Digital Image Processing*; Pearson Education India: Chennai, India, 2009.
40. Gedraite, E.S.; Hadad, M. Investigation on the effect of a Gaussian Blur in image filtering and segmentation. In Proceedings of the ELMAR-2011, Zadar, Croatia, 14–16 September 2011; pp. 393–396.
41. Kumar, N.; Nachamai, M. Noise removal and filtering techniques used in medical images. *Orient. J. Comput. Sci. Technol* **2017**, *10*, 103–113. [CrossRef]
42. Gençtav, A.; Aksoy, S.; Önder, S. Unsupervised segmentation and classification of cervical cell images. *Pattern Recognit.* **2012**, *45*, 4151–4168. [CrossRef]
43. Mir, M.A.; Qazi, F.; Naseem, M.; Zia, S.S.; Mubeen, T. Invisibility Cloak using Color Extraction and Image Segmentation with OpenCV. In Proceedings of the 2022 Global Conference on Wireless and Optical Technologies (GCWOT), Malaga, Spain, 14–17 February 2022; pp. 1–6.
44. Luo, Y.; Duraiswami, R. Canny edge detection on NVIDIA CUDA. In Proceedings of the 2008 IEEE Computer Society Conference on Computer Vision and Pattern Recognition Workshops, Anchorage, AL, USA, 23–28 June 2008; pp. 1–8.
45. Punagin, A.; Punagin, S. Analysis of lane detection techniques on structured roads using OpenCV. *Int. J. Res. Appl. Sci. Eng. Technol.* **2020**, *8*, 2994–3003. [CrossRef]
46. Piccinini, F.; Peirsman, A.; Stellato, M.; Pyun, J.c.; Tumedei, M.M.; Tazzari, M.; De Wever, O.; Tesei, A.; Martinelli, G.; Castellani, G. Deep learning-based tool for morphotypic analysis of 3D multicellular spheroids. *J. Mech. Med. Biol.* **2023**, *23*, 2340034. [CrossRef]
47. Stellato, M.; Rydzyk, M.M.; Pannella, M.; Rossi, F.; Cappadone, C.; Remondini, D.; Pyun, J.C.; Normanno, N.; Ibrahim, T.; Castellani, G.; et al. Radiomic analysis of 3D spheroids using 2D brightfield images. *Biomed. Signal Process. Control.* **2025**, *103*, 107366. [CrossRef]
48. McKenzie, A.J.; Hicks, S.R.; Svec, K.V.; Naughton, H.; Edmunds, Z.L.; Howe, A.K. The mechanical microenvironment regulates ovarian cancer cell morphology, migration, and spheroid disaggregation. *Sci. Rep.* **2018**, *8*, 7228. [CrossRef]
49. Zhang, Q.; Lin, F.; Huang, J.; Xiong, C. Mechanical transmission enables EMT cancer cells to drive epithelial cancer cell migration to guide tumor spheroid disaggregation. *Sci. China Life Sci.* **2022**, *65*, 2031–2049. [CrossRef] [PubMed]
50. Burleson, K.M.; Boente, M.P.; Pambuccian, S.E.; Skubitz, A.P. Disaggregation and invasion of ovarian carcinoma ascites spheroids. *J. Transl. Med.* **2006**, *4*, 1–16. [CrossRef]
51. Nawrocki Raby, B.; Polette, M.; Gilles, C.; Clavel, C.; Strumane, K.; Matos, M.; Zahm, J.M.; Van Roy, F.; Bonnet, N.; Birembaut, P. Quantitative cell dispersion analysis: New test to measure tumor cell aggressiveness. *Int. J. Cancer* **2001**, *93*, 644–652. [CrossRef]
52. Pinto, B.; Henriques, A.C.; Silva, P.M.; Bousbaa, H. Three-dimensional spheroids as in vitro preclinical models for cancer research. *Pharmaceutics* **2020**, *12*, 1186. [CrossRef] [PubMed]
53. Piccinini, F.; Tesei, A.; Arienti, C.; Bevilacqua, A. Cancer multicellular spheroids: Volume assessment from a single 2D projection. *Comput. Methods Programs Biomed.* **2015**, *118*, 95–106. [CrossRef] [PubMed]
54. Song, Y.; Chen, J.; Zhang, C.; Xin, L.; Li, Q.; Liu, Y.; Zhang, C.; Li, S.; Huang, P. Mechanosensitive channel Piezo1 induces cell apoptosis in pancreatic cancer by ultrasound with microbubbles. *iScience* **2022**, *25*, 103733. [CrossRef] [PubMed]

applied sciences

MDPI

Article

Effects of Varying Caffeine Dosages and Consumption Timings on Cerebral Vascular and Cognitive Functions: A Diagnostic Ultrasound Study

Min-Ki Choi [1,†], Hee-Seul Ahn [1,†], Da-Eun Kim [1], Da-Seul Lee [1], Chan-Sol Park [2,*] and Chang-Ki Kang [1,2,3,*]

[1] Department of Radiological Science, College of Medical Science, Gachon University, Incheon 21936, Republic of Korea
[2] Department of Health Science, Gachon University Graduate School, Incheon 21936, Republic of Korea
[3] Neuroscience Research Institute, Gachon University, Incheon 21565, Republic of Korea
* Correspondence: cs11111kr@gachon.ac.kr (C.-S.P.); ckkang@gachon.ac.kr (C.-K.K.)
† These authors contributed equally to this work.

Abstract: Caffeine is consumed owing to its stimulatory effects; however, its excessive intake triggers adverse effects. Herein, we analyzed changes in physiological cerebrovascular and cognitive functions following the consumption of 100 and 200 mg of caffeine in healthy adults after 0/30/60 min to ascertain appropriate caffeine consumption habits. The peak systolic velocity (PSV), pulse wave velocity (PWV), and common carotid artery (CCA) diameter were measured using diagnostic ultrasound. Cognitive function was evaluated using the accuracy rate and response time on the two-back task. Percutaneous oxygen saturation (SpO_2) and heart rate (HR) were assessed using patient monitoring systems. After consuming 100 mg of caffeine, systolic blood pressure (SBP) increased ($p > 0.05$) and SpO_2 and accuracy rate improved by 30 min ($p = 0.018$ and $p = 0.356$) but declined by 60 min ($p = 0.924$ and $p = 0.055$). HR and response time continuously decreased ($p = 0.209$ and $p = 0.061$, respectively), while PWV showed no change ($p > 0.05$). After consuming 200 mg of caffeine, SBP ($p < 0.05$), diastolic blood pressure ($p = 0.004$ and $p = 0.820$), and SpO_2 ($p = 0.002$ and $p = 0.666$) increased significantly, while the accuracy rate ($p = 0.634$ and $p = 0.055$, respectively) and response time ($p < 0.05$) decreased. PWV remained unchanged ($p > 0.05$). The results revealed distinct dose-dependent patterns on physiological and cognitive changes, with SBP and SpO_2 exhibiting greater changes when a higher dose was consumed in a short duration. Although moderate caffeine intake may support vascular health and cognitive function, excessive intake reduces blood flow, alters vascular elasticity, and impairs cognitive activation. These findings highlight the need for guidelines to ensure safe and effective caffeine consumption.

Keywords: caffeine dose; cognitive function; common carotid artery; diagnostic ultrasound; pulse wave velocity

1. Introduction

Caffeine is consumed worldwide in various forms such as coffee, chocolate, beverages, and pharmaceuticals [1]. The caffeine content varies by form, with approximately 17 mg/oz in coffee, 0.8 mg/oz in cocoa and hot chocolate, and 30–100 mg in medications [2]. Most caffeine intake occurs through coffee, which is consumed at various times throughout the day, such as after waking up or during meals [1,3]. Globally, since 2002, the coffee consumption per capita has increased by 37% over the last 20 years [4], resulting in a paralleled increase in caffeine intake. Currently, the U.S. FDA recommends a maximum daily caffeine intake of 400 mg for healthy adults. Similarly, Health Canada recommends maximum daily doses of 45–85 mg/day for children aged <12, 2.5 mg/kg/day for adolescents aged >12 years, 400 mg/day for adults, and 300 mg/day for pregnant women [5]. The European Food Safety Authority (EFSA) recommends daily caffeine intake limits of 0.2–2.0 mg/kg body weight (bw)/day for children (3–10 years), 0.4–1.4 mg/kg bw/day for adolescents (10–18 years), 400 mg/day for adults, and 200 mg/day for pregnant women. The Ministry of Food and Drug Safety of South Korea recommends maximum daily intakes of 2.5 mg/kg bw/day for children and adolescents, 400 mg/day for adults, and 300 mg/day for pregnant women [6,7]. Despite these guidelines, public awareness remains low, resulting in excessive caffeine consumption without consideration of the recommended limits. The primary driver of this consumption is the stimulatory effect of caffeine.

Caffeine exerts its stimulatory effects through multiple mechanisms. Its structural similarity to adenosine allows it to act as an antagonist, with stimulatory effects being a key outcome [8]. Adenosine typically binds to the A1 and A2A receptors in the brain, thereby reducing the release of glutamate and dopamine and inducing drowsiness and sleepiness [9]. However, caffeine binds to these receptors non-selectively, blocking adenosine-mediated signaling and enhancing neurotransmitter release. This results in reduced drowsiness, improved brain function, and sympathetic nervous system activation, which increases catecholamine release and alertness [8,10,11]. Additionally, caffeine induces vasoconstriction, elevates blood pressure (BP) [8,12,13], increases renal blood flow, and inhibits ion and fluid reabsorption, leading to diuretic effects [8]. After exerting these effects, caffeine is metabolized by the liver and excreted by the kidneys through the urine [14,15]. Its half-life in the body averages 3–5 h and varies according to physiological and environmental factors [8,16,17].

The positive effects of caffeine extend beyond its stimulatory effects. It activates lipase, promotes fat breakdown [14,15,18], enhances muscle strength by affecting muscle contraction [8,15,19], and aids in digestion by stimulating gastrin release and gastric acid secretion in the stomach and duodenum [20,21]. However, excessive caffeine intake can cause dehydration due to excessive diuresis, gastritis, or ulcers due to the overproduction of gastric acid [8,21], in addition to psychological side effects such as anxiety, tension, and insomnia [22]. High accessibility to caffeine also affects adolescent consumption, potentially causing electrolyte imbalances that impair growth and development, with potentially adverse effects on mental health during critical periods of physical and emotional change [23,24]. In pregnant women, excessive caffeine intake may result in caffeine crossing the placental barrier owing to its lipid solubility, potentially causing miscarriages or other complications [25].

The effects of caffeine, both in terms of dosage and duration, are well documented. While small doses of caffeine may improve cognitive function and vascular health, excessive amounts can lead to reduced blood flow, an increased heart rate (HR), and impaired cognitive performance [18–20]. Previous studies have reported cardiovascular responses, such as increased BP and HRs within 30–60 min after caffeine consumption, as well as improvements in cerebral oxygen saturation and working memory [21,22]. However, most studies have focused on specific doses (>300 mg) and have not systematically analyzed dose-dependent differences or combined vascular and cognitive changes over time.

To address the above knowledge gaps, the present study aimed to analyze the physiological and cognitive responses to varying doses of caffeine intake in a short duration using diagnostic ultrasonography and cognitive function task.

2. Materials and Methods

2.1. Participants

This study was approved by the Institutional Review Board (IRB No. 1044396-202404-HR-067-01). The participants in this study were recruited through bulletin boards of recruiting target institutions or public announcements posted on social networks, and 30 participants (male: 15; female: 15) voluntarily participated (Table 1). Prior to the experiment, participants were informed about this study's purpose, procedures, and equipment safety. All participants provided their written informed consent to participate in this study.

Table 1. Participant information (mean ± SD).

	Age (Years)	Height (cm)	Weight (kg)	BMI (kg/m^2)
Males (n = 15)	22.93 ± 1.33	174.93 ± 6.25	74.73 ± 11.27	24.35 ± 2.94
Females (n = 15)	22.53 ± 1.36	159.72 ± 3.22	53.38 ± 5.62	20.89 ± 1.75
Total (n = 30)	22.73 ± 1.33	167.33 ± 9.15	64.06 ± 13.94	22.63 ± 2.96

Abbreviations: BMI, body mass index; SD, standard deviation.

The criteria for inclusion were healthy adults over the age of 20 without any disease, and the exclusion criteria are those who have a history of diseases possibly affecting cardiovascular function, take cardiovascular drugs, are physically weak or mentally ill, and are students affiliated with the same department as the researchers.

2.2. Experimental Protocols and Data Acquisition

The participants completed two experimental sessions held at the same time, one week apart. To minimize external factors influencing the results, the participants refrained from smoking and eating for 6 h before each session, avoided caffeine consumption for 12 h, and abstained from alcohol consumption for 24 h. The 12 h caffeine restriction was set according to its half-life: approximately 3–5 h.

During the measurements, all participants were seated on a fixed chair in the same position, with the back of their heads touching the wall. Velcro tape and a polyester band were placed around their foreheads to secure their heads in place and prevent any movement away from the wall.

A diagnostic ultrasound system (E-CUBE i7, ALPINION MEDICAL SYSTEMS Co., Ltd., Seoul, Republic of Korea) and a linear transducer array (L3-12T, ALPINION MEDICAL SYSTEMS Co., Ltd., Seoul, Republic of Korea) were used to locate and mark the right CCA for consistent measurements. The ultrasound measurements were performed by a radiologic technologist with experience in multiple ultrasound-related studies. BP was monitored using a smartwatch (SM-R850; Samsung Electronics, Suwon, Republic of Korea) worn on the left wrist and calibrated individually according to the manufacturer's instructions. The HR and peripheral oxygen saturation (SpO_2) were measured using a patient monitor (Bionics, Chuncheon, Republic of Korea), with sensors attached to the left index finger.

To assess cognitive performance, the participants practiced the N-back test once before the experiment to familiarize themselves with the task. The two-back (2B) test was then conducted using the DMDX software (DMDX version 6, University of Arizona Psychology Department, Tucson, AZ, USA). In this test, numbers ranging from 0 to 9 appeared randomly on the screen 40 times at 3 s intervals. Participants were required to indicate whether the current number matched the one presented two trials earlier by selecting "positive" or "negative."

Subjective measures were evaluated using the visual analog scale (VAS) for comfort, Karolinska Sleepiness Scale (KSS) for drowsiness, and the Chalder Fatigue Scale (CFS) for fatigue. The VAS ranges from the most comfortable state (0) to the most uncomfortable state (10), whereas the KSS ranges from highest alertness (1) to greatest sleepiness (9) [26,27]. The CFS uses a 14-item Likert scale (eight items for physical fatigue and six for mental fatigue), with responses scored as better than usual (0), same as usual (1), worse than usual (2), and much worse than usual (3), yielding a total score of 0–42, with higher scores indicating greater fatigue [28].

Physiological and biosignal measurements were taken at three time points: before caffeine ingestion (0 min), 30 min after ingestion (30 min), and 60 min after ingestion (60 min), based on caffeine's approximate absorption time of 45 min in the bloodstream [29]. Brightness mode (B-mode) and Doppler-mode ultrasound images of the right CCA were recorded for 5 s, capturing eight cardiac cycles (Figure 1). Blood pressure (BP) was measured concurrently using a smartwatch.

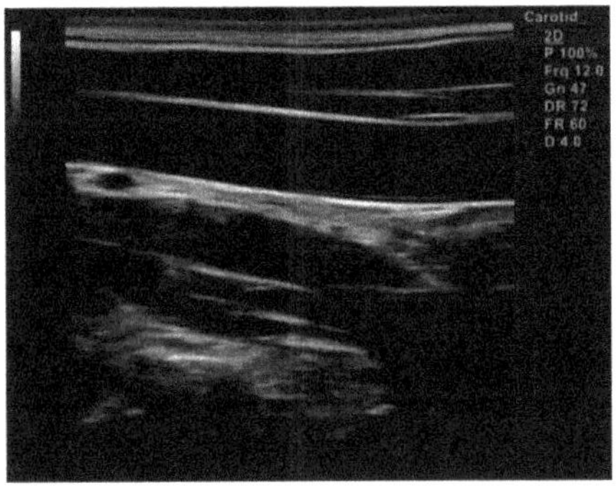

Figure 1. Brightness mode image of CCA measured using ultrasound.

Participants consumed caffeine immediately following the 0 min measurement period. Caffeine doses were prepared using stick coffee (Kanu Mini Mild-Roasted Coffee Sticks; Dongsuh Foods Corporation, Incheon, Republic of Korea), containing 33 mg of caffeine per stick. For each trial, three (approximately 100 mg) or six sticks (approximately 200 mg) were dissolved in 200 mL of water. In the first experiment, the participants were randomly assigned to consume either 100 mg or 200 mg of caffeine, and the alternate dose was consumed in the second experiment. Therefore, a single-blind method was used, in which the participants were unaware of the intake amount.

Ultrasound images of the CCA were saved in DICOM format and analyzed using DICOM viewer software (Radiant DICOM Viewer Version 2021.2.2 (64-bit), Medixant, Poznan, Poland). The maximum diameter (MaxD) and minimum diameter (MinD) were measured from the B-mode images, while the average peak systolic velocity (PSV) over 5 s was calculated from the Doppler-mode images. HR and SpO_2 values were averaged over 30 s of measurement, and systolic blood pressure (SBP) and diastolic blood pressure (DBP) were recorded using the smartwatch application.

The pulse wave velocity (PWV), a measure of arterial stiffness, was calculated using SBP, DBP, MaxD, and MinD according to Equations (1) and (2). Cognitive performance data, including accuracy rate and response time from the 2B test, were extracted using the DMDX software.

$$\text{Stiffness Index (SI)} = \ln\left(\frac{\text{SBP}}{\text{DBP}}\right) \times \left(\frac{\text{MinD}}{\Delta\text{D}}\right) \tag{1}$$

where DD = MaxD − MinD

$$\text{Pulse Wave Velocity (PWV)} = \sqrt{\frac{\text{SI} \times \text{DBP}}{2 \times \text{BD}}} \quad (2)$$

where blood density (BD) = 1.050 g/cm^3

The raw data for BP, diameter, PWV, PSV, HR, SpO_2, accuracy rate, and response time were normalized using Equation (3) to calculate the change rates for statistical analysis.

$$\text{Change Rate (CR)} = \frac{\textit{Value at } 0, 30, \textit{or } 60 \text{ min}}{\textit{Value at } 0 \text{ min}} \quad (3)$$

2.3. Statistical Analysis

Statistical analyses were conducted using Jamovi software (version 2.2.5, free software, https://www.jamovi.org, accessed on 15 October 2024). Time-dependent changes (0, 30, and 60 min) were analyzed using repeated-measures analysis of variance (RM ANOVA). Sphericity was tested, and Greenhouse–Geisser corrections were applied where necessary. Tukey's post hoc test was performed for significant effects ($p < 0.05$). Changes in caffeine dose (100 mg vs. 200 mg) were analyzed using paired *t*-tests. For subjective measures, the means and standard deviations were calculated to observe trends according to caffein doses and consumption times. Statistical significance was set at $p = 0.05$.

3. Results

All measurements were expressed as relative change rates, with the baseline at 0 min set to 1. Overall, SBP showed no significant differences at 30 (1.003 ± 0.037) or 60 min (1.009 ± 0.030) following the intake of 100 mg of caffeine ($p > 0.05$). However, following a 200 mg intake, SBP significantly increased at 30 (1.014 ± 0.029) and 60 min (1.029 ± 0.044) ($p < 0.05$) (Table 2).

Table 2. Comparison of blood pressure parameters using RM ANOVA.

Variation	Caffeine Dose	Time	Mean ± SD	F	p	Factor 1	Factor 2	Mean Difference ± SE (Factor 1–Factor 2)	t	p_{Tukey}
SBP	100 mg	0 min	1	1.4	0.254	0 min	30 min	−0.00301 ± 0.00669	−0.450	0.895
		30 min	1.003 ± 0.03666				60 min	−0.00921 ± 0.00552	−1.667	0.235
		60 min	1.009 ± 0.03026			30 min	60 min	−0.00620 ± 0.00436	−1.420	0.344
	200 mg	0 min	1	10.3	<0.001 *	0 min	30 min	−0.01400 ± 0.00536	−2.610	0.037 *
		30 min	1.014 ± 0.02936				60 min	−0.02930 ± 0.00804	−3.640	0.003 *
		60 min	1.029 ± 0.04403			30 min	60 min	−0.01530 ± 0.00563	−2.720	0.029 *
DBP	100 mg	0 min	1	0.179	0.836	0 min	30 min	−0.00221 ± 0.00866	−0.255	0.965
		30 min	1.002 ± 0.04777				60 min	0.00262 ± 0.00842	0.311	0.948
		60 min	0.997 ± 0.04639			30 min	60 min	0.00483 ± 0.00706	0.685	0.774
	200 mg	0 min	1	7.85	0.001 *	0 min	30 min	−0.02122 ± 0.00595	−3.564	0.004 *
		30 min	1.021 ± 0.03207				60 min	−0.02551 ± 0.00752	−3.390	0.006 *
		60 min	1.022 ± 0.03649			30 min	60 min	−0.00429 ± 0.00711	−0.603	0.820

Abbreviations: DBP, diastolic blood pressure; F, F-statistic; *p*, *p*-value; SBP, systolic blood pressure; SE, standard Error; t, t-statistic. * $p < 0.05$

DBP showed no significant differences at 30 (1.002 ± 0.048) or 60 min (0.997 ± 0.046) following 100 mg of caffeine intake ($p > 0.05$), whereas for the 200 mg intake, significant increases were observed from 0 to 30 min and 0 to 60 min ($p = 0.004$ and $p = 0.006$, respectively); however, no significant differences were noted between 30 and 60 min ($p = 0.820$) (Table 2 and Figure 2).

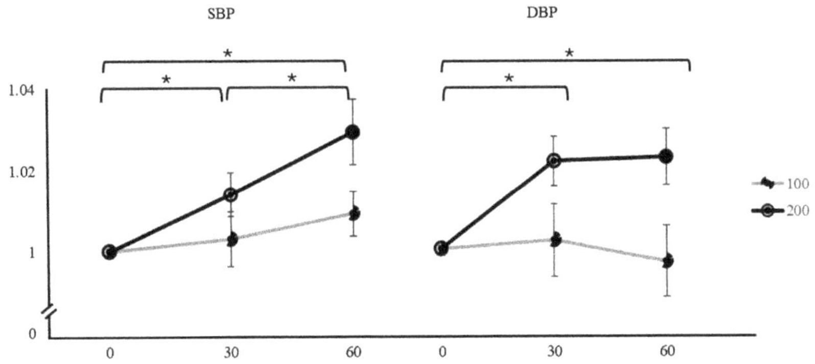

Figure 2. Comparison of changes in SBP and DBP at each time point (the error bars indicate standard error). * $p < 0.05$.

Significant differences were observed between 0 and 30 min ($p = 0.016$), but no significant differences were observed between 0 and 60 or 30 and 60 min ($p = 0.504$ and $p = 0.138$, respectively) following 100 mg of caffeine intake. For the 200 mg intake, no significant differences were observed at 30 min (1.001 ± 0.028) or 60 min (0.999 ± 0.030) ($p > 0.05$) (Table 3).

Table 3. Comparison of diameter using RM ANOVA.

Variation	Caffeine Dose	Time	Mean ± SD	F	*p*	Factor 1	Factor 2	Mean Difference ± SE (Factor 1–Factor 2)	t	p_{Tukey}
MaxD	100 mg	0 min	1	4.610	0.014 *	0 min	30 min	−0.01777 ± 0.00594	−2.990	0.016 *
		30 min	1.020 ± 0.02951				60 min	−0.0071 ± 0.00628	−1.130	0.504
		60 min	1.008 ± 0.03317			30 min	60 min	0.01067 ± 0.00541	1.970	0.138
	200 mg	0 min	1	0.130	0.878	0 min	30 min	−0.00070 ± 0.00504	−0.136	0.990
		30 min	1.001 ± 0.02800				60 min	0.00214 ± 0.00549	0.389	0.920
		60 min	0.999 ± 0.02998			30 min	60 min	0.00282 ± 0.00665	0.424	0.906
MinD	100 mg	0 min	1	4.840	0.011 *	0 min	30 min	−0.01902 ± 0.00707	−2.689	0.031 *
		30 min	1.020 ± 0.03820				60 min	−0.01298 ± 0.00520	−2.497	0.048 *
		60 min	1.013 ± 0.02849			30 min	60 min	0.00604 ± 0.00632	0.956	0.610
	200 mg	0 min	1	0.427	0.655	0 min	30 min	0.00229 ± 0.00674	0.340	0.938
		30 min	0.999 ± 0.03647				60 min	0.00545 ± 0.00571	0.953	0.612
		60 min	0.996 ± 0.03033			30 min	60 min	0.00315 ± 0.00520	0.606	0.818

Abbreviations: F, F-statistic; MaxD, maximum diameter; MinD, minimum diameter; *p*, *p*-value; SE, standard error; t, t-statistic. * $p < 0.05$

MinD showed significant differences from 0 to 30 min and 0 to 60 min ($p = 0.031$ and $p = 0.048$, respectively), but no significant differences were found between 30 and 60 min ($p = 0.610$) after 100 mg of caffeine intake. For the 200 mg intake, no significant differences were observed from 30 min (0.999 ± 0.036) and 60 min (0.996 ± 0.030) ($p > 0.05$) (Table 3 and Figure 3).

The PWV showed no significant differences at 30 (1.005 ± 0.136) or 60 min (1.053 ± 0.112) following 100 mg of caffeine intake ($p > 0.05$). For the 200 mg intake, no significant differences were observed at 30 (1.014 ± 0.162) or 60 min (1.001 ± 0.128) ($p > 0.05$) (Table 4 and Figure 4).

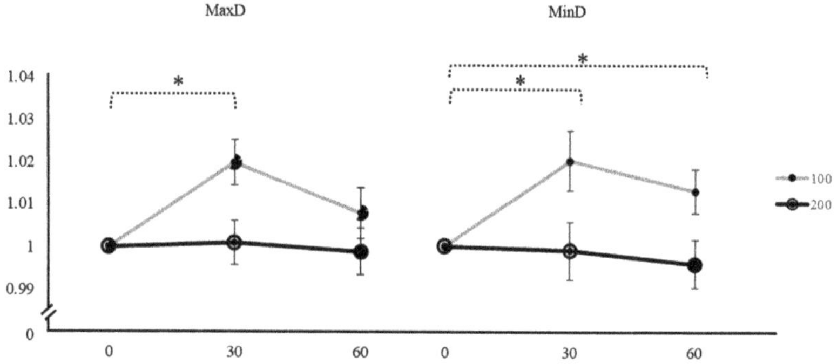

Figure 3. Comparison of changes in the MaxD and MinD of the CCA at each time point (the error bars indicate standard error). * $p < 0.05$.

Table 4. Comparison of PWV using RM ANOVA.

Variation	Caffeine Dose	Time	Mean ± SD	F	p	Factor 1	Factor 2	Mean Difference ± SE (Factor 1–Factor 2)	t	p_{Tukey}
PWV	100 mg	0 min	1	2.860	0.066	0 min	30 min	−0.00245 ± 0.0254	−0.0966	0.995
		30 min	1.005 ± 0.13648				60 min	−0.05081 ± 0.0208	−2.4374	0.055
		60 min	1.053 ± 0.11166			30 min	60 min	−0.04836 ± 0.0253	−1.9090	0.156
	200 mg	0 min	1	0.695	0.503	0 min	30 min	−0.02844 ± 0.0313	−0.9090	0.639
		30 min	1.014 ± 0.16218				60 min	−0.00319 ± 0.0229	−0.1390	0.989
		60 min	1.001 ± 0.12751			30 min	60 min	0.02525 ± 0.0243	1.0390	0.559

Abbreviations: F, F-statistic; p, p-value; PWV, pulse wave velocity; SE, standard error; t, t-statistic.

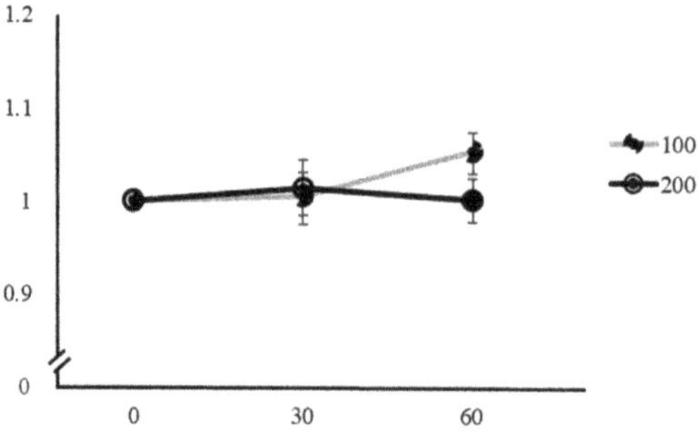

Figure 4. Comparison of changes in the PWV of the CCA at each time point (the error bars indicate standard error).

The PSV showed significant differences between 0 and 30 min ($p < 0.001$); however, no significant differences were observed between 0 and 60 min and 30 and 60 min ($p = 0.056$ and $p = 0.407$, respectively) following 100 mg of caffeine intake. For the 200 mg intake, no significant differences were observed from 30 min (0.960 ± 0.069) or 60 min (0.961 ± 0.099) ($p > 0.05$) (Table 5 and Figure 5).

HR showed significant differences between 0 and 60 min ($p = 0.005$) but no significant differences between 0 and 30 min and 30 and 60 min ($p = 0.209$ and $p = 0.061$, respectively) after 100 mg of caffeine intake. For the 200 mg intake group, significant differences were only observed from 0 to 60 min ($p = 0.022$), with insignificant changes observed between 0 and 30 and 30 and 60 min ($p = 0.080$, $p = 0.235$, respectively) (Table 5 and Figure 6).

Table 5. Comparison of the PSV, HR, and SpO_2 using RM ANOVA.

Variation	Caffeine Dose	Time	Mean ± SD	F	*p*	Factor 1	Factor 2	Mean Difference ± SE (Factor 1–Factor 2)	t	p_{Tukey}
PSV	100 mg	0 min	1	7.98	<0.001 *	0 min	30 min	0.0768 ± 0.0185	4.140	<0.001 *
		30 min	0.927 ± 0.10060				60 min	0.0526 ± 0.0217	2.420	0.056
		60 min	0.956 ± 0.11457			30 min	60 min	−0.0242 ± 0.0186	−1.300	0.407
	200 mg	0 min	1	2.11	0.131	0 min	30 min	0.02918 ± 0.0150	1.951	0.144
		30 min	0.960 ± 0.06945				60 min	0.03403 ± 0.0189	1.797	0.190
		60 min	0.961 ± 0.09885			30 min	60 min	0.00484 ± 0.0195	0.248	0.967
HR	100 mg	0 min	1	8.04	0.001 *	0 min	30 min	0.0112 ± 0.0064	1.760	0.209
		30 min	0.988 ± 0.03054				60 min	0.0310 ± 0.00875	3.540	0.005 *
		60 min	0.973 ± 0.03853			30 min	60 min	0.0198 ± 0.00814	2.430	0.061
	200 mg	0 min	1	6.09	0.008 *	0 min	30 min	0.0187 ± 0.00830	2.250	0.080
		30 min	0.980 ± 0.04016				60 min	0.0298 ± 0.01049	2.840	0.022 *
		60 min	0.974 ± 0.05511			30 min	60 min	0.0111 ± 0.00662	1.670	0.235
SpO_2	100 mg	0 min	1	5.28	0.008 *	0 min	30 min	−0.0043 ± 0.00148	−2.914	0.018 *
		30 min	1.004 ± 0.00808				60 min	−0.0038 ± 0.00153	−2.482	0.049 *
		60 min	1.003 ± 0.00808			30 min	60 min	0.0005 ± 0.00133	0.379	0.924
	200 mg	0 min	1	11.5	<0.001 *	0 min	30 min	−0.00693 ± 0.00181	−3.821	0.002 *
		30 min	1.006 ± 0.00913				60 min	−0.00834 ± 0.00211	−3.947	0.001 *
		60 min	1.008 ± 0.01113			30 min	60 min	−0.00142 ± 0.00164	−0.866	0.666

Abbreviations: F, F-statistic; HR, heart rate; PSV, peak systolic velocity; *p*, *p*-value; SE, standard error; SpO_2, peripheral oxygen saturation; t, t-statistic. * $p < 0.05$

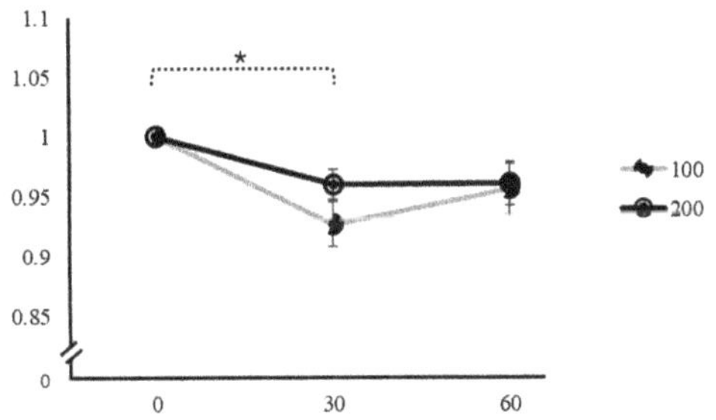

Figure 5. Comparison of changes in the PSV of the CCA at each time point (the error bars indicate standard error). * $p < 0.05$.

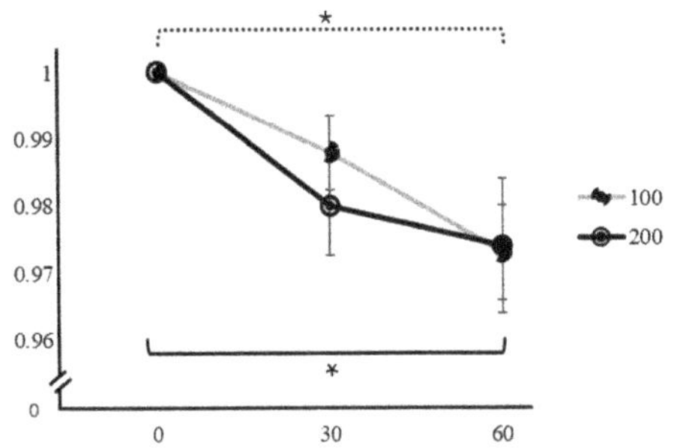

Figure 6. Comparison of changes in HR at each time point (the error bars indicate standard error). * $p < 0.05$.

SpO_2 showed significant differences between 0 and 30 min and 0 and 60 min ($p = 0.018$ and $p = 0.049$, respectively) but no significant differences between 30 and 60 min ($p = 0.924$)

after 100 mg of caffeine intake. For the 200 mg intake, significant differences were observed from 0 to 30 min and 0 to 60 min ($p = 0.002$ and $p = 0.001$, respectively) but not between 30 and 60 min ($p = 0.666$) (Table 5 and Figure 7).

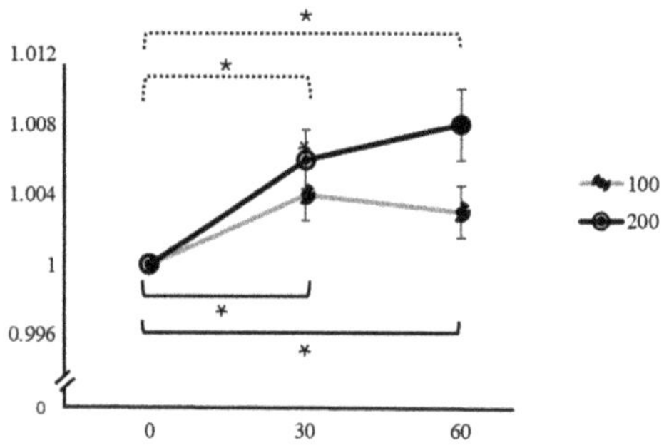

Figure 7. Comparison of changes in SpO_2 at each time point (the error bars indicate standard error). * $p < 0.05$.

The accuracy rate significantly differed between 30 and 60 min ($p = 0.009$); however, no significant differences were observed between 0 and 30 min and 0 and 60 min ($p = 0.356$ and $p = 0.613$, respectively) following 100 mg of caffeine intake. For the 200 mg intake, significant differences were observed between 0 and 60 min ($p = 0.020$) but not observed from 0 to 30 and 30 to 60 min ($p = 0.634$ and $p = 0.055$, respectively) (Table 6 and Figure 8).

Table 6. Comparison of results of the two-back test using RM ANOVA.

Variation	Caffeine Dose	Time	Mean ± SD	F	p	Factor 1	Factor 2	Mean Difference ± SE (Factor 1–Factor 2)	t	p_{Tukey}
Accuracy rate	100 mg	0 min	1	3.31	0.044 *	0 min	30 min	−0.01305 ± 0.00932	−1.400	0.356
		30 min	1.013 ± 0.04842				60 min	0.00903 ± 0.00947	0.953	0.613
		60 min	0.991 ± 0.04923			30 min	60 min	0.02207 ± 0.00684	3.226	0.009 *
	200 mg	0 min	1	5.44	0.007 *	0 min	30 min	0.00537 ± 0.00586	0.917	0.634
		30 min	0.995 ± 0.03218				60 min	0.02034 ± 0.00708	2.874	0.020 *
		60 min	0.979 ± 0.04009			30 min	60 min	0.01497 ± 0.00618	2.423	0.055
Response time	100 mg	0 min	1	30.4	<0.001 *	0 min	30 min	0.0445 ± 0.0094	4.740	<0.001 *
		30 min	0.963 ± 0.04238				60 min	0.0701 ± 0.01048	6.690	<0.001 *
		60 min	0.934 ± 0.04590			30 min	60 min	0.0256 ± 0.00706	3.620	0.003 *
	200 mg	0 min	1	32.6	<0.001 *	0 min	30 min	0.0232 ± 0.00660	3.520	0.005 *
		30 min	0.976 ± 0.03408				60 min	0.0637 ± 0.00928	6.860	<0.001 *
		60 min	0.938 ± 0.04713			30 min	60 min	0.0404 ± 0.00784	5.160	<0.001 *

Abbreviations: F, F-statistic; p, p-value; SE, standard error; t, t-statistic. * $p < 0.05$

The response time showed significant differences over time following both the 100 mg and 200 mg caffeine intake, with reductions observed between 30 min (0.963 ± 0.042, 0.976 ± 0.034) and 60 min (0.934 ± 0.046, 0.938 ± 0.047) ($p < 0.05$) (Table 6 and Figure 9).

Caffeine dosage significantly affected SBP at 60 min ($p = 0.016$) but not at 30 min ($p = 0.145$) (Table 7). Similarly, DBP significantly differed at 60 min ($p = 0.031$) but not at 30 min ($p = 0.059$) (Table 7). Furthermore, MaxD and MinD exhibited significant differences at 30 min ($p = 0.019$ and $p = 0.029$, respectively) but not at 60 min ($p = 0.052$) (Table 7). PWV, PSV, and HR showed no significant dose-dependent effects over time ($p > 0.05$), whereas the SpO_2 differences were significant at 60 min ($p < 0.05$) but not at 30 min ($p = 0.338$) (Table 7). The accuracy rate and response time were not significantly affected ($p > 0.05$) (Table 7).

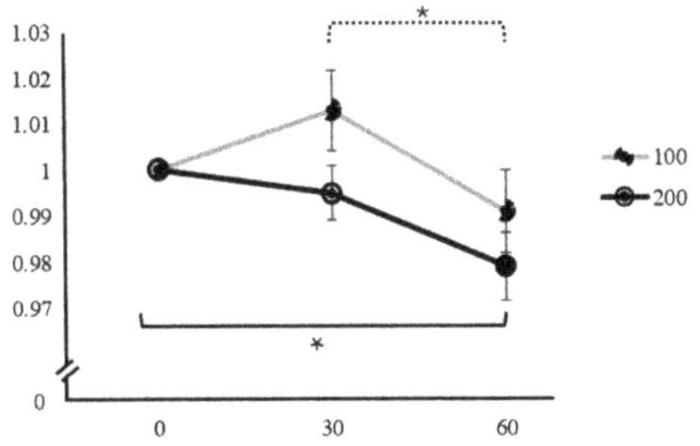

Figure 8. Comparison of the changes in accuracy rate at each time point (the error bars indicate standard error). * $p < 0.05$.

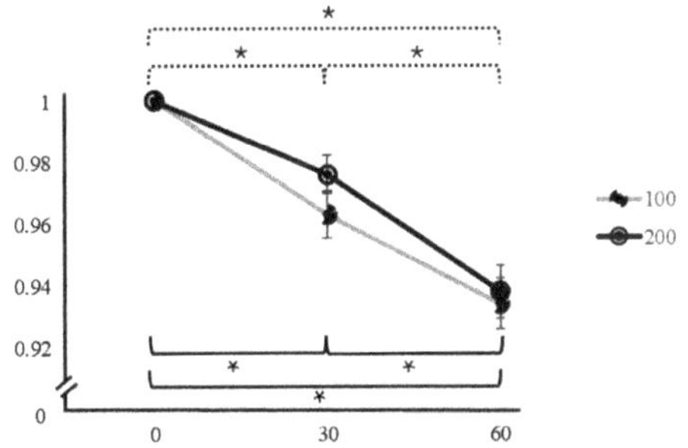

Figure 9. Comparison of the changes in response time at each time point (the error bars indicate standard error). * $p < 0.05$.

Table 7. Comparison of the change rate by caffeine dose using paired *t*-test.

Variation	Time	Mean Difference ± SE (100 mg–200 mg)	t	*p*
SBP	30 min	−0.01097 ± 0.00732	−1.4985	0.145
	60 min	−0.02008 ± 0.00786	−2.5530	0.016 *
DBP	30 min	−0.01852 ± 0.00939	−1.9720	0.059
	60 min	−0.02498 ± 0.01097	−2.2760	0.031 *
MaxD	30 min	0.01953 ± 0.00783	2.495	0.019 *
	60 min	0.00837 ± 0.00638	1.311	0.201
MinD	30 min	0.02147 ± 0.00930	2.307	0.029 *
	60 min	0.01718 ± 0.00844	2.035	0.052
PWV	30 min	−0.00904 ± 0.04381	−0.0206	0.838
	60 min	0.05229 ± 0.03464	1.5100	0.143
PSV	30 min	−0.03298 ± 0.02186	−1.509	0.144
	60 min	−0.00489 ± 0.03181	−0.154	0.879
HR	30 min	0.00818 ± 0.00897	0.912	0.373
	60 min	−0.00185 ± 0.01311	−0.141	0.889
SpO_2	30 min	−0.002 ± 0.00205	−0.976	0.338
	60 min	−0.00437 ± 0.00213	−2.051	0.050 *
Accuracy rate	30 min	0.01782 ± 0.01168	1.526	0.139
	60 min	0.01242 ± 0.01464	0.849	0.404
Response time	30 min	−0.01296 ± 0.01064	−1.218	0.235
	60 min	−0.00403 ± 0.0117	−0.357	0.724

Abbreviations: DBP, diastolic blood pressure; HR, heart rate; MaxD, maximum diameter; MinD, minimum diameter; PSV, peak systolic velocity; PWV, pulse wave velocity; *p*, *p*-value; SBP, systolic blood pressure; SE, standard Error; SpO_2, peripheral oxygen saturation; t, t-statistic. * $p < 0.05$

At a caffeine intake of 100 mg, the VAS score was 3.13 ± 2.19 at 0 min, 3.07 ± 2.03 at 30 min, and 2.80 ± 2.04 at 60 min. This value decreased by 0.06 at 30 min compared to that at 0 min and by 0.27 at 60 min compared to that at 30 min. For a caffeine intake of 200 mg, the VAS was 2.87 ± 2.16 at 0 min, 2.83 ± 2.09 at 30 min, and 2.57 ± 2.05 at 60 min. This value decreased by 0.03 at 30 min compared to that at 0 min and by 0.26 at 60 min compared to that at 30 min.

For a caffeine intake of 100 mg, the KSS was 4.87 ± 1.59 at 0 min, 4.00 ± 1.46 at 30 min, and 4.13 ± 1.41 at 60 min. This value decreased by 0.46 at 30 min compared to 0 min and by 0.28 at 60 min compared to 30 min. When the caffeine intake was 200 mg, the KSS was 5.00 ± 1.55 at 0 min, 3.87 ± 1.22 at 30 min, and 3.87 ± 1.28 at 60 min. The 200 mg intake decreased by 1.13 at 30 min compared to that at 0 min and showed no further decrease from 30 min to 60 min (0.00 change).

For a caffeine intake of 100 mg, the CFS was 16.30 ± 5.53 at 0 min, 14.40 ± 5.62 at 30 min, and 16.00 ± 5.64 at 60 min. This value decreased by 1.90 at 30 min compared to that at 0 min and increased by 2.10 at 60 min compared to that at 30 min. For a caffeine intake of 200 mg, the CFS was 14.80 ± 5.58 at 0 min, 13.20 ± 5.64 at 30 min, and 12.80 ± 5.31 at 60 min. This value decreased by 1.60 at 30 min compared to that at 0 min and decreased further by 0.40 at 60 min compared to that at 30 min (Table 8).

Table 8. Results of subjective evaluation.

Variation	Caffeine Dose	0 min	Δ	30 min	Δ	60 min
VAS	100 mg	3.13 ± 2.19	−0.06	3.07 ± 2.03	−0.27	2.8 ± 2.04
	200 mg	2.87 ± 2.16	−0.03	2.83 ± 2.09	−0.26	2.57 ± 2.05
KSS	100 mg	4.87 ± 1.59	−0.46	4.00 ± 1.46	−0.28	4.13 ± 1.41
	200 mg	5.00 ± 1.55	−1.13	3.87 ± 1.22	0.00	3.87 ± 1.28
CFS	100 mg	16.30 ± 5.53	−1.90	14.40 ± 5.62	2.10	16.00 ± 5.64
	200 mg	14.80 ± 5.58	−1.60	13.20 ± 5.64	−0.40	12.80 ± 5.31

Abbreviations: CFS, Chalder Fatigue Scale; KSS, Karolinska Sleepiness Scale; VAS, visual analog scale.

In short, the VAS scores decreased slightly at 60 min compared to those at 0 min for both doses, although this difference was insignificant. The KSS decreased at 30 min and showed minor increases at 60 min, indicating reduced sleepiness following caffeine intake. The CFS scores showed a transient reduction at 30 min with a slight rebound at 60 min for 100 mg and continued reduction at 200 mg (Table 8).

4. Discussion

In this study, the effects of caffeine intake on BP, HR, SpO_2, blood vessel diameter, and cognitive function were analyzed. The BP change was not significant when 100 mg was ingested in a short duration, but BP increased significantly after 200 mg of intake. HR tended to decrease at both doses, and SpO_2 increased further when 200 mg was ingested. In the cognitive function evaluation, the accuracy improved until 30 min after 100 mg of intake but decreased thereafter, and the accuracy continuously decreased when 200 mg was ingested. On the other hand, the reaction time was shortened under all conditions, suggesting the possibility that caffeine may have a positive effect on rapid cognitive response.

SBP demonstrated a general upward trend over time across all levels of caffeine intake, with a significant increase observed only at 200 mg. This increase could be attributed to the stimulation of the sympathetic nervous system, leading to catecholamine release, which induces vasoconstriction and subsequent BP elevation [8,10,13,20]. The most pronounced increase occurred approximately 45 min post-intake, aligning with the peak absorption of caffeine [29,30]. DBP showed a modest increase for up to 30 min following 100 mg of intake, followed by a decrease at 60 min. Conversely, a 200 mg intake resulted in a gradual increase in DBP up to 60 min. These observations are consistent with the mechanisms underlying

SBP changes, indicating that the limited increase in DBP after 30 min of consuming 200 mg of caffein may reflect a physiological threshold influenced by blood flow. The changes in SBP and DBP were more pronounced with 200 mg than with 100 mg, with significant differences observed only at 60 min, indicating that higher caffeine doses exert greater effects on BP.

The MaxD and MinD increased at 30 min compared to the baseline with a 100 mg intake but decreased by 60 min. Following an intake of 200 mg, MaxD slightly increased at 30 min and subsequently slightly decreased at 60 min, while MinD progressively declined. These findings suggest that high doses of caffeine have a strong effect on vasoconstriction. Changes in MaxD and MinDs were more pronounced at 100 mg than at 200 mg, with significant differences observed only at 30 min, indicating a shorter recovery period for vessel diameter compared to BP [9,13].

PWV remained relatively stable at 30 min for all doses. At 100 mg, it increased slightly at 60 min, likely due to the smooth muscle contraction triggered by elevated intracellular calcium levels, resulting in vascular stiffening and increased wave propagation speed [31,32]. In contrast, the PWV at 200 mg remained constant for up to 60 min. Neither time nor dosage resulted in significant differences in the PWV. Previous studies have reported either the maintenance of or an increase in PWV following acute caffeine intake [32,33]. Differences in measurement methods, such as foot-to-foot versus the one-point approach, may account for these discrepancies. The latter provides more localized measurements, thereby reducing variability.

Overall, PSV exhibited a decreasing trend 30 min after intake across all doses. Recovery was observed after 60 min for 100 mg, whereas a slight decrease was observed for 200 mg. Although the reduction observed at 30 min was more substantial at a 100 mg caffein dose, both doses showed similar levels at 60 min, without significant differences. This aligns with the effects of caffeine on the carotid and peripheral arteries, where vasoconstriction impedes smooth blood flow owing to reduced adenosine activity [8,34,35]. HR decreased significantly over time for all doses, with greater reductions at 200 mg than at 100 mg at 30 min. After 60 min, both of the doses resulted in similar reductions. This trend is likely a baroreflex response to elevated BP, aimed at reflexively lowering HR [36–38]. SpO_2 increased across all doses up to 30 min, followed by a slight decline at 100 mg, and continued to increase at 200 mg up to 60 min. These results are consistent with previous findings indicating that a low dose of caffeine reduces tissue oxygen consumption, leading to increased SpO_2 levels [39,40].

In cognitive function tests, comprising the 2_B test assessment, the accuracy rate increased at 30 min for 100 mg but returned to baseline by 60 min, whereas it showed a continuous decline until 60 min after a 200 mg dose. The transient improvement at 30 min with the lower dose reflected the peak absorption and immediate effects of caffeine [29,30]. However, the subsequent decline highlights the short-lived nature of the cognitive effects of caffeine. Increased sympathetic activity due to higher caffeine doses likely contributed to the reduced accuracy. The participants' stationary posture during the experiment may have activated the default mode network, which is associated with reduced focus and concentration [41–46]. The response time consistently improved over time for all doses, indicating enhanced cognitive arousal with caffeine intake, regardless of the dose.

Subjective measures, such as the VAS, KSS, and CFS, demonstrated increased comfort and reduced sleepiness and fatigue at 30 and 60 min for both doses. VAS and KSS showed greater changes at 100 mg, whereas CFS showed greater changes at 200 mg. These findings indicate the arousal effects of caffeine, which are mediated by dopaminergic modulation in the prefrontal cortex, which enhances mood and relaxation [47–49]. The general expectation of the effects of caffeine may have influenced the subjective evaluations regardless of the actual dose [50].

Overall, our findings indicate that caffeine exerts both dose- and time-dependent effects on physiological and cognitive parameters. Moderate caffeine intake appears to be beneficial for maintaining vascular health and cognitive activation, while excessive intake may lead to adverse vascular changes and cognitive impairment.

The limitations of this study include that the participants of the experiment were limited to their 20 s, so it was not possible to compare the results according to age, and the effect of caffeine on the patients with any disease could not be confirmed. In addition, the change after 60 min could not be confirmed because the measurement was made within a limited time up to 60 min after caffeine intake. The number of study participants is still small. Therefore, in a future study, it is necessary to supplement these limitations and increase the number of samples to conduct studies on various age groups and patients with diseases and additional studies on changes after 60 min.

Unlike some previous studies that were conducted in the lying down position and/or with a single dose [32,51], this study was conducted in a sitting position and with two different doses. In addition, this study examined the vascular physiology of caffeine, cognitive function, and subjective changes in comparison to the previous study [51].

5. Conclusions

Caffeine, which induces physiological and cognitive changes, is widely consumed in various forms, such as in coffee, chocolate, and medications. This study quantitatively analyzed the effects of caffeine intake (100 mg and 200 mg) and elapsed time (0, 30, and 60 min) on vascular physiology, cognitive function, and other subjective measures. The results revealed no significant changes in blood pressure or cognitive function with a 100 mg dose. However, the 200 mg dose significantly increased systolic blood pressure and diastolic blood pressure and impaired cognitive accuracy. Subjective measurements indicated reduced sleepiness and fatigue across all doses, reflecting the arousal effects of caffeine. This study highlighted the importance of moderately consuming caffeine. Low-dose caffeine intake over a short period is relatively safe and supports vascular and cognitive health, whereas high-dose caffeine intake over a short period causes vascular constriction and cognitive decline.

The practical implications of this study are as follows. Consuming a moderate amount of caffeine can promote vascular health and cognitive function. In addition, individualized guidelines should account for varying physiological responses to caffeine. Lastly, strategic use of the positive effects of caffeine, coupled with measures to prevent excessive consumption, is crucial.

Future research should explore the effects of caffeine in diverse age groups, health conditions, and the long-term effects of caffeine consumption to establish comprehensive guidelines.

Author Contributions: M.-K.C., H.-S.A., D.-E.K., D.-S.L., C.-S.P. and C.-K.K. were involved in conceptualization, data curation, formal analysis, investigation, methodology, resources, software, validation, visualization, and writing the original draft. C.-S.P. and C.-K.K. were involved in project administration, writing, reviewing, and editing as corresponding authors. C.-K.K. was involved in supervision and funding acquisition. All authors have read and agreed to the published version of the manuscript.

Funding: This research was supported by grants from the National Research Foundation of Korea (NRF), funded by the Korean government (MSIT) (No. 2020R1A2C1004355; No. 2022R1F1A1062766).

Institutional Review Board Statement: The study was conducted in accordance with the Declaration of Helsinki, and approved by the Institutional Review Board of Gachon university (1044396-202404-HR-067-01, approval on 21 June 2024).

Informed Consent Statement: Informed consent was obtained from all participants.

Data Availability Statement: The data will be made available to interested individuals upon reasonable request.

Conflicts of Interest: The authors declare no conflicts of interest.

References

1. Nieber, K. The impact of coffee on health. *Planta Med.* **2017**, *83*, 1256–1263. [CrossRef] [PubMed]
2. Barone, J.J.; Roberts, H.R. Caffeine consumption. *Food Chem. Toxicol.* **1996**, *34*, 119–129. [CrossRef] [PubMed]
3. Kang, Y.-J. Effects of caffeine on nerve conduction velocity. *J. Converg. Inf. Technol.* **2020**, *10*, 195–199. [CrossRef]
4. Quadra, G.R.; Brovini, E.M.; Dos Santos, J.A.; Paranaíba, J.R. Caffeine consumption over time. In *Handbook of Substance Misuse and Addictions*; Patel, V.B., Preedy, V.R., Eds.; Springer International Publishing: Cham, Switzerland, 2022; pp. 1535–1552; ISBN 978-3-030-92391-4.
5. Mitchell, D.C.; Knight, C.A.; Hockenberry, J.; Teplansky, R.; Hartman, T.J. Beverage caffeine intakes in the U.S. *Food Chem. Toxicol.* **2014**, *63*, 136–142. [CrossRef]
6. EFSA Panel on Dietetic Products, Nutrition and Allergies. Scientific Opinion on the safety of caffeine. *EFS2* **2015**, *13*, 4102. [CrossRef]
7. Lee, J.-S.; Park, H.-S.; Han, S.; Tana, G.; Chang, M.-J. Study on relationship between caffeine intake level and metabolic syndrome and related diseases in Korean adults: 2013~2016 Korea National Health and Nutrition Examination Survey. *J. Nutr. Health* **2019**, *52*, 227. [CrossRef]
8. Reddy, V.S.; Shiva, S.; Manikantan, S.; Ramakrishna, S. Pharmacology of caffeine and its effects on the human body. *Eur. J. Med. Chem. Rep.* **2024**, *10*, 100138. [CrossRef]
9. Layland, J.; Carrick, D.; Lee, M.; Oldroyd, K.; Berry, C. Adenosine: Physiology, pharmacology, and clinical applications. *JACC Cardiovasc. Intv.* **2014**, *7*, 581–591. [CrossRef]
10. McLellan, T.M.; Caldwell, J.A.; Lieberman, H.R. A review of caffeine's effects on cognitive, physical and occupational performance. *Neurosci. Biobehav. Rev.* **2016**, *71*, 294–312. [CrossRef]
11. Davis, J.M.; Zhao, Z.; Stock, H.S.; Mehl, K.A.; Buggy, J.; Hand, G.A. Central nervous system effects of caffeine and adenosine on fatigue. *Am. J. Physiol. Regul. Integr. Comp. Physiol.* **2003**, *284*, R399–R404. [CrossRef]
12. Cappelletti, S.; Piacentino, D.; Sani, G.; Aromatario, M. Caffeine: Cognitive and physical performance enhancer or psychoactive drug? *Curr. Neuropharmacol.* **2015**, *13*, 71–88. [CrossRef] [PubMed]
13. Echeverri, D.; Montes, F.R.; Cabrera, M.; Galán, A.; Prieto, A. Caffeine's vascular mechanisms of action. *Int. J. Vasc. Med.* **2010**, *2010*, 834060. [CrossRef] [PubMed]
14. Zulli, A.; Smith, R.M.; Kubatka, P.; Novak, J.; Uehara, Y.; Loftus, H.; Qaradakhi, T.; Pohanka, M.; Kobyliak, N.; Zagatina, A.; et al. Caffeine and cardiovascular diseases: Critical review of current research. *Eur. J. Nutr.* **2016**, *55*, 1331–1343. [CrossRef] [PubMed]
15. Tavares, C.; Sakata, R.K. Caffeine in the treatment of pain. *Rev. Bras. Anestesiol.* **2012**, *62*, 387–401. [CrossRef] [PubMed]
16. Turnbull, D.; Rodricks, J.V.; Mariano, G.F.; Chowdhury, F. Caffeine and cardiovascular health. *Regul. Toxicol. Pharmacol.* **2017**, *89*, 165–185. [CrossRef]
17. Reich, N.; Mannino, M.; Kotler, S. Using caffeine as a chemical means to induce flow states. *Neurosci. Biobehav. Rev.* **2024**, *159*, 105577. [CrossRef]
18. Da Silva, L.A.; Pereira, R.A.; Túrmina, J.A.; Kerppers, I.I.; Osiecki, R.; Altimari, L.R.; Malfatti, C.R.M. Sulfonylurea induction of caffeine-enhanced insulin secretion and reduction of glycemic levels in diabetic rats. *Pharm. Biol.* **2014**, *52*, 956–960. [CrossRef]
19. Bakker, R.; Steegers, E.A.P.; Obradov, A.; Raat, H.; Hofman, A.; Jaddoe, V.W.V. Maternal caffeine intake from coffee and tea, fetal growth, and the risks of adverse birth outcomes: The generation R Study. *Am. J. Clin. Nutr.* **2010**, *91*, 1691–1698. [CrossRef]
20. Ruxton, C.H.S. The impact of caffeine on mood, cognitive function, performance and hydration: A review of benefits and risks. *Nutr. Bull.* **2008**, *33*, 15–25. [CrossRef]
21. Nehlig, A. Effects of coffee on the gastro-intestinal tract: A narrative review and literature update. *Nutrients* **2022**, *14*, 399. [CrossRef]
22. Delleli, S.; Ouergui, I.; Messaoudi, H.; Trabelsi, K.; Ammar, A.; Glenn, J.M.; Chtourou, H. Acute effects of caffeine supplementation on physical performance, physiological responses, perceived exertion, and technical-tactical skills in combat sports: A systematic review and meta-analysis. *Nutrients* **2022**, *14*, 2996. [CrossRef] [PubMed]
23. Chang, Y.-E.; Chung, H.-K. Survey of caffeine intake from children's favorite foods. *Korean J. Nutr.* **2010**, *43*, 475. [CrossRef]

24. Kim, N.; Shin, W.; Kim, Y. Study on relevance of high-caffeine drink intake frequency to mental health of adolescents. *Korean J. Food Culture* **2017**, *32*, 66–74. [CrossRef]
25. Qian, J.; Chen, Q.; Ward, S.M.; Duan, E.; Zhang, Y. Impacts of caffeine during pregnancy. *Trends Endocrinol. Metab.* **2020**, *31*, 218–227. [CrossRef]
26. Geiger Brown, J.; Wieroney, M.; Blair, L.; Zhu, S.; Warren, J.; Scharf, S.M.; Hinds, P.S. Measuring subjective sleepiness at work in hospital nurses: Validation of a modified delivery format of the Karolinska sleepiness scale. *Sleep Breath.* **2014**, *18*, 731–739. [CrossRef]
27. Kaida, K.; Takahashi, M.; Åkerstedt, T.; Nakata, A.; Otsuka, Y.; Haratani, T.; Fukasawa, K. Validation of the Karolinska sleepiness scale against performance and EEG variables. *Clin. Neurophysiol.* **2006**, *117*, 1574–1581. [CrossRef]
28. Seo, B.N.; Kim, I. Differences of stress, sleep quality and metabolic syndrome by prolonged fatigue in early adulthood. *Korean J. Adult Nurs.* **2019**, *31*, 202. [CrossRef]
29. Alsabri, S.G.; Mari, W.O.; Younes, S.; Elsadawi, M.A.; Oroszi, T.L. Kinetic and dynamic description of caffeine. *J. Caffeine Adenosine Res.* **2018**, *8*, 3–9. [CrossRef]
30. Mort, J.R.; Kruse, H.R. Timing of blood pressure measurement related to caffeine consumption. *Ann. Pharmacother.* **2008**, *42*, 105–110. [CrossRef]
31. Sato, K.; Sanders, K.M.; Gerthoffer, W.T.; Publicover, N.G. Sources of calcium utilized in cholinergic responses in canine colonic smooth muscle. *Am. J. Physiol.* **1994**, *267*, C1666–C1673. [CrossRef]
32. Mahmud, A.; Feely, J. Acute effect of caffeine on arterial stiffness and aortic pressure waveform. *Hypertension* **2001**, *38*, 227–231. [CrossRef] [PubMed]
33. Echeverri, D.; Pizano, A.; Montes, F.R.; Forcada, P. Acute effect of coffee consumption on arterial stiffness, evaluated using an oscillometric method. *Artery Res.* **2017**, *17*, 16. [CrossRef]
34. Dix, L.M.L.; van Bel, F.; Baerts, W.; Lemmers, P.M.A. Effects of caffeine on the preterm brain: An observational study. *Early Hum. Dev.* **2018**, *120*, 17–20. [CrossRef] [PubMed]
35. Ozkan, B.; Yüksel, N.; Anik, Y.; Altintas, O.; Demirci, A.; Çağlar, Y. The effect of caffeine on retrobulbar hemodynamics. *Curr. Eye Res.* **2008**, *33*, 804–809. [CrossRef] [PubMed]
36. McClaran, S.R.; Wetter, T.J. Low doses of caffeine reduce heart rate during submaximal cycle ergometry. *J. Int. Soc. Sports Nutr.* **2007**, *4*, 11. [CrossRef]
37. Sullivan, J.J.; Knowlton, R.G.; Brown, D.D. Caffeine affects heart rate and blood pressure response to prolonged walking. *J. Cardiopulm. Rehabil.* **1992**, *12*, 418–422. [CrossRef]
38. Mosqueda-Garcia, R.; Tseng, C.J.; Biaggioni, I.; Robertson, R.M.; Robertson, D. Effects of caffeine on baroreflex activity in humans. *Clin. Pharmacol. Ther.* **1990**, *48*, 568–574. [CrossRef]
39. Ruíz-Moreno, C.; Lara, B.; Brito De Souza, D.; Gutiérrez-Hellín, J.; Romero-Moraleda, B.; Cuéllar-Rayo, Á.; Del Coso, J. Acute caffeine intake increases muscle oxygen saturation during a maximal incremental exercise test. *Br. J. Clin. Pharmacol.* **2020**, *86*, 861–867. [CrossRef]
40. Gonzaga, L.A.; VanderLei, L.C.M.; Gomes, R.L.; Valenti, V.E. Caffeine affects autonomic control of heart rate and blood pressure recovery after aerobic exercise in young adults: A crossover study. *Sci. Rep.* **2017**, *7*, 14091. [CrossRef]
41. Menon, V. 20 years of the default mode network: A review and synthesis. *Neuron* **2023**, *111*, 2469–2487. [CrossRef]
42. Luo, W.; Liu, B.; Tang, Y.; Huang, J.W.; Wu, J. Rest Promotes Learning: From the Perspective of Brain's Default Mode Network Perspective. *Behav. Sci.* **2024**, *14*, 349. [CrossRef] [PubMed]
43. Cheng, X.; Yuan, Y.; Wang, Y.; Wang, R. Neural antagonistic mechanism between default-mode and task-positive networks. *Neurocomputing* **2020**, *417*, 74–85. [CrossRef]
44. Devaney, K.J.; Levin, E.J.; Tripathi, V.; Higgins, J.P.; Lazar, S.W.; Somers, D.C. Attention and default mode network assessments of meditation experience during active cognition and rest. *Brain Sci.* **2021**, *11*, 566. [CrossRef] [PubMed]
45. Bonnelle, V.; Leech, R.; Kinnunen, K.M.; Ham, T.E.; Beckmann, C.F.; De Boissezon, X.; Greenwood, R.J.; Sharp, D.J. Default mode network connectivity predicts sustained attention deficits after traumatic brain injury. *J. Neurosci.* **2011**, *31*, 13442–13451. [CrossRef]
46. Koshino, H.; Minamoto, T.; Yaoi, K.; Osaka, M.; Osaka, N. Coactivation of the default mode network Regions and Working Memory Network regions during task preparation. *Sci. Rep.* **2014**, *4*, 5954. [CrossRef]
47. Astorino, T.A.; Roberson, D.W. Efficacy of acute caffeine ingestion for short-term high-intensity exercise performance: A systematic review. *J. Strength Cond. Res.* **2010**, *24*, 257–265. [CrossRef]
48. Alasmari, F. Caffeine induces neurobehavioral effects through modulating neurotransmitters. *Saudi Pharm. J.* **2020**, *28*, 445–451. [CrossRef]
49. Subramaniam, K.; Vinogradov, S. Improving the neural mechanisms of cognition through the pursuit of happiness. *Front. Hum. Neurosci.* **2013**, *7*, 452. [CrossRef]

50. Beedie, C.J. All in the mind? Pain, placebo effect, and ergogenic effect of caffeine in sports performance. *Open Access J. Sports Med.* **2010**, *1*, 87–94. [CrossRef]
51. Jin, Y.-B.; Kim, J.-H.; Song, C.-H.; Park, C.; Kang, C.-K. Diagnostic Ultrasound-Based Investigation of Central vs. Peripheral Arterial Changes Consequent to Low-Dose Caffeine Ingestion. *Nutrients* **2024**, *16*, 228. [CrossRef]

applied sciences

MDPI

Review

Focused Ultrasound (FUS) and Pediatric Brain Tumors: Current Status and Future Directions

Sarah Kleinknecht [1,*], Kristen Fox [2], Fotios Tsitsos [3] and Stergios Zacharoulis [1,*]

[1] Division of Hematology-Oncology, Department of Pediatrics, NewYork-Presbyterian Morgan Stanley Children's Hospital, Columbia University Vagelos College of Physicians and Surgeons, New York, NY 10032, USA
[2] Department of Pediatrics, NewYork-Presbyterian Morgan Stanley Children's Hospital, Columbia University Vagelos College of Physicians and Surgeons, New York, NY 10032, USA; waf9015@nyp.org
[3] Department of Biomedical Engineering, Columbia University, New York, NY 10032, USA; ft2561@columbia.edu
* Correspondence: sk5432@cumc.columbia.edu (S.K.); sz2764@cumc.columbia.edu (S.Z.)

Abstract: Diffuse intrinsic pontine glioma (DIPG), or as it is newly redefined, diffuse midline glioma (DMG), remains one of the most horrific diagnoses in pediatric oncology. Aggressive and inaccessible to standard treatments, it is generally considered incurable. Focused ultrasound technology has developed over the last several decades as a noninvasive means to target various types of tumors in both adults and children. Recent advances, particularly in low-intensity focused ultrasound (LIFU), have opened new avenues for enhancing drug delivery and modulating the tumor microenvironment in these challenging tumors. This review provides a comprehensive overview of preclinical and clinical research developments in the use of LIFU for pediatric DMGs. We highlight key findings from animal models demonstrating improved blood–brain barrier (BBB) permeability, increased chemotherapeutic and nanoparticle delivery, and potential immunomodulatory effects of LIFU. Emerging clinical studies, including early-phase safety and feasibility trials, are also discussed, with attention to technical parameters, imaging guidance strategies, and biomarkers of response. The review concludes by addressing the challenges of translating LIFU into routine clinical practice, including device optimization for pediatric anatomy, regulatory hurdles, and the need for standardized treatment protocols. Collectively, these recent advances underscore the promise of LIFU as a minimally invasive, image-guided adjunct to current and future therapies for pediatric DMGs, warranting continued research and collaborative clinical efforts.

Keywords: diffuse midline glioma; focused ultrasound; pediatric oncology; drug delivery; blood–brain barrier

1. Introduction

Brain tumors are the second most common childhood malignancy and the most common solid tumors in children, as well as the leading cause of cancer-related mortality [1]. Among pediatric central nervous system (CNS) tumors, diffuse midline gliomas account for approximately 10% to 15% and occur at an incidence of 0.35 per 100,000 children per year [2,3]. These tumors are WHO grade 4 gliomas that arise in the pons, thalamus and other midline structures, and they most commonly affect children between the ages of 5 and 9 [3].

Historically, treatment strategies in neuro-oncology have centered around three primary modalities: surgery, radiation therapy, and chemotherapy. Although progress in molecular classification and subtyping has significantly advanced our understanding of tumor biology and informed the development of targeted therapies, effective treatment remains limited. One of the major barriers to therapeutic advancement is the presence of the blood–brain barrier, which severely restricts drug delivery to the central nervous system [4].

Focused ultrasound (FUS) is a noninvasive therapeutic modality with a range of applications in both malignant and non-malignant conditions. While its clinical use has expanded in adult populations, it received FDA approval for use in pediatric patients only in 2020 [5]. FUS encompasses several approaches, including high-intensity focused ultrasound (HIFU), low-intensity focused ultrasound (LIFU), and sonodynamic therapy, all of which non-invasively transmit acoustic energy across the skull. These techniques rely on the transmission of mechanical energy at specific frequencies to target tissue, with differing biological effects based on frequency and intensity [6]. HIFU, typically operating at 650 kHz, induces thermal ablation of targeted tissue, whereas LIFU at lower frequencies (e.g., 220 kHz) can modulate neuronal activity and transiently open the BBB, particularly when combined with microbubble contrast agents (Figure 1) [1].

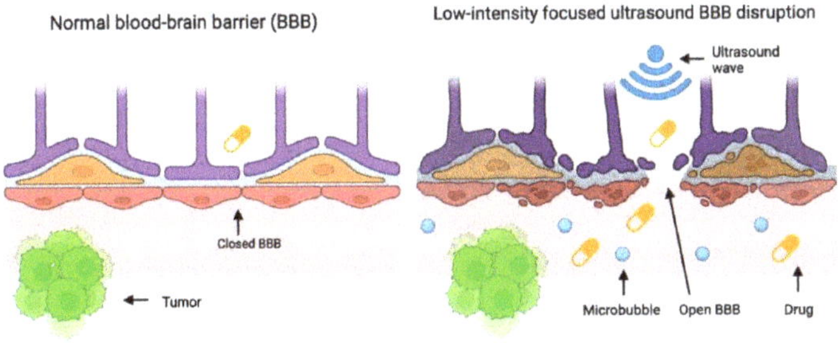

Figure 1. Mechanism of low-intensity focused ultrasound.

The ability of LIFU to transiently and reversibly disrupt the BBB presents a promising strategy to enhance drug delivery to infiltrative CNS tumors such as DMGs. Over the past two decades, numerous preclinical studies have demonstrated the feasibility and safety of this approach, with increasing evidence supporting its utility in enhancing delivery of chemotherapeutic agents, antibodies, and nanoparticle-based therapies to brain tumors. More recently, this work has begun to translate into early-phase clinical trials, reflecting growing interest in its potential clinical application.

This review provides an up-to-date overview of recent and ongoing preclinical and clinical studies using LIFU in the treatment of pediatric diffuse intrinsic pontine glioma/diffuse midline glioma, highlighting its evolving role in the neuro-oncologic treatment landscape and outlining current challenges and future directions.

2. Methodology

A scoping review was performed in accordance with the Preferred Reporting Items for Systematic Reviews and Meta-Analyses Extension for Scoping Reviews (PRISMA-ScR) guidelines (Figure 2). To evaluate past and present preclinical studies regarding focused ultrasound, a comprehensive literature review was conducted using publications retrieved from PubMed using the keywords "low-intensity focused ultrasound", "pediatric DIPG focused ultrasound" and "LIFU DMG". Articles included peer-reviewed original research and technical reports regarding experiments which examined FUS-mediated BBB disruption in DIPG/DMG animal models during the time period 2021–2025. A systematic review was also performed of active clinical studies from clinicaltrials.gov. Clinical trials that were excluded were those involving sonodynamic therapy as the focus of this review was low-intensity focused ultrasound. Subsequent investigation of any notable results of these trials thus far was ascertained through further PubMed search individualized to each NCT identification number.

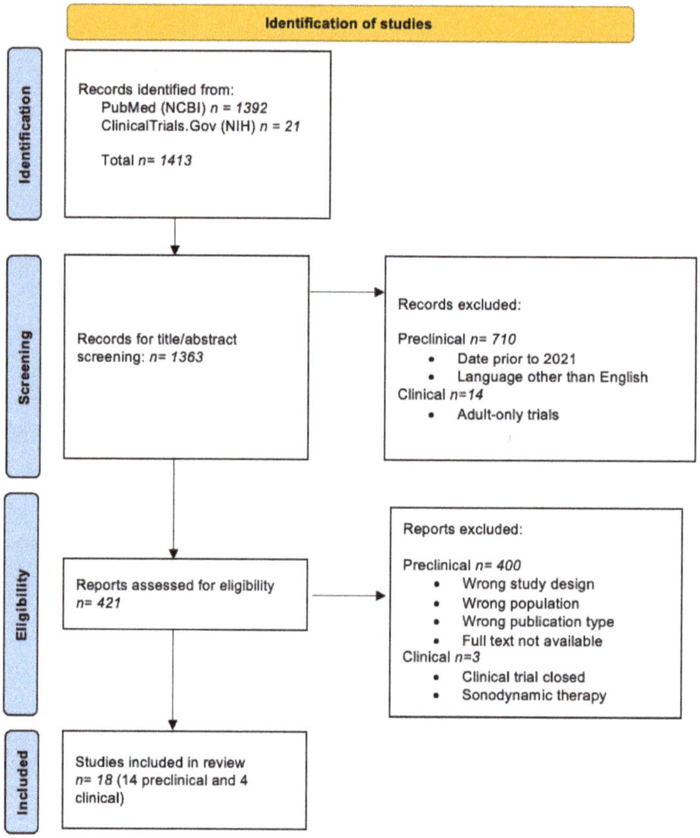

Figure 2. PRISMA flowchart for study identification.

From the selected preclinical and clinical sources, the following was collected for each: study design and model type, FUS modality (LIFU), therapeutic adjunct (chemotherapy, sonosensitizers), technical parameters, safety outcomes, feasibility, technical challenges, as well as early efficacy or biological endpoints, where available.

3. Results

In total, 14 preclinical analyses and four clinical trials of FUS in pediatric DMG in the selected time frame were found. DIPG models demonstrated overall that FUS-mediated BBB opening significantly enhances various therapeutic agent accumulation within tumors, for a wide variety of different chemotherapies, radiation, immunotherapies, and biologic modifiers. And several ongoing early-phase clinical trials are evaluating the safety and feasibility of these approaches in pediatric patients with DMG. Of note, due to high heterogeneity of the included studies, this review mainly carries out narrative synthesis rather than meta-analysis.

4. Preclinical Studies Evaluating LIFU and DMG

4.1. Chemotherapy

Doxorubicin

Doxorubicin, a compound of the anthracycline class, is well known as an extremely powerful antineoplastic drug with one of the broadest spectrums of activity, and has been well-studied for use in DIPG [7]. An initial preclinical study by Sewing et al. demonstrated that high-grade glioma and DIPG cells exhibit sensitivity to doxorubicin, while normal human astrocytes are relatively spared [8]. However, its clinical use in central nervous system tumors has been limited by its inability to cross the BBB due to its large size and hydrophilicity [5]. To circumvent these limitations, Filieri et al. highlighted the rationale for FUS-mediated BBB disruption to facilitate localized doxorubicin delivery [9].

In 2021, Kim et al. used a closed-loop, trans-skull MR-guided FUS hyperthermia system optimized at ~1.7 MHz to deliver controlled, localized heating though the skull, minimizing off-target effects [10]. Safety experiments showed that hyperthermia at 41.5 ± 0.5 °C for 10 min could be maintained consistently without tissue injury. Using this data, the group tested thermosensitive liposomal doxorubicin (TSL-Dox) in GL261 glioma-bearing mice and F98 glioma-bearing rats. Hyperthermia was induced using FUS at 42 °C for 10–20 min post-injection. They found that FUS-mediated hyperthermia significantly improved drug delivery, with about a 3.5-fold increase in doxorubicin uptake in mice and a 5-fold increase in rats compared to TSL-Dox alone. Fluorescence microscopy confirmed that in the absence of hyperthermia, doxorubicin remained localized to the vessel wall, while FUS heating promoted deeper penetration into tumor tissue. To investigate this mechanism, the investigators used physiologically based pharmacokinetic modeling and dynamic contrast-enhanced MRI which both demonstrated that thermal stress increased vascular permeability and transport capacity, with higher permeability values observed in FUS-treated tumors. These results indicated that the benefit of hyperthermia came from both the triggered release of doxorubicin and the changes in tumor vascularity. Importantly, in a survival study conducted with the GL261-bearing mice, the combination of TSL-Dox and FUS hyperthermia led to a significant extension of median survival compared to the TSL-Dox group alone.

Subsequently, Choi et al. investigated a novel two-step FUS protocol in male Sprague Dawley rats implanted with intracranial 9L gliosarcoma tumors [11]. Their design included three study groups: FUS without microbubbles, FUS with microbubble-mediated BBB disruption, and a combined sequential strategy consisting of FUS stimulation followed by microbubble-mediated blood–brain barrier disruption. Using MRI guidance, researchers found that the sequential approach significantly increased blood-tumor permeability and doxorubicin delivery compared to either method alone. T1-weighted contrast-enhanced MRI showed a 2.65-fold increase in signal intensity in the tumor region compared to the contralateral control, and a 1.45-fold increase compared to microbubble-mediated BBB disruption. Dynamic contrast-enhanced MRI further demonstrated a 2.08-fold increase in permeability in tumor regions, which was also 1.25-fold higher than BBB disruption alone. In terms of drug accumulation, the sequential protocol produced a 1.91-fold increase in doxorubicin concentration within the tumor compared to control tissue, outperforming the 1.44-fold increase seen with BBB disruption alone. Importantly, histopathological examination revealed no evidence of tissue injury across groups, supporting the safety of this multi-step approach.

4.2. Temozolomide (TMZ)

Already commonly utilized in the treatment of pediatric brain tumors, due to its bioavailability, ability to cross the BB, and induction of cytotoxicity, TMZ makes for an intriguing combination with FUS. However, its efficacy is frequently affected by resistance mechanisms, particularly overexpression of the DNA repair enzyme O6-methylguanine-DNA methyltransferase (MGMT), which directly reverses TMZ-induced DNA lesions and thereby allows tumors to survive despite treatment. To overcome this issue, Yang et al. developed lipid-polymer hybrid nanoparticles (LPHNs) functionalized with cyclic RGD peptides (cRGD) for targeted delivery to tumors with CRISPR/Cas9 plasmids targeting the MGMT gene in orthotopic glioblastoma-bearing nude mice implanted with human T98G glioblastoma cells [12]. FUS with microbubbles was then used to transiently and safely open the BBB, enabling efficient nanoparticle delivery into the tumor microenvironment. Imaging studies confirmed that this approach enhanced intratumoral drug accumulation. The triple combination of FUS, CRISPR-loaded LPHNs, and TMZ produced robust tumor suppression, smaller tumor volumes on serial MRI, and a marked improvement in survival, extending median lifespan to 43 days compared with 22–30 days in other groups.

4.3. Epirubicin

Epirubicin, another derivative of the anthracycline class, is hindered in clinical application for CNS tumors by the restrictive nature of the BBB. To address this limitation, Shen et al. created a DNA nanocarrier loaded with epirubicin and tested its delivery in combination with FUS and microbubble-mediated BBB disruption [13]. The epirubicin loaded DNA nanocarriers were injected into Sprague Dawley rats implanted with C6 rat glioma cells and FUS was applied to target tumor regions. Quantitative fluorescence imaging and histopathologic analysis confirmed a 4.4-fold increase in drug accumulation within tumor cells. Additionally, tumor growth was suppressed by nearly 87% at day 23 compared to controls, and survival was significantly prolonged, with median survival extended to about 36.5 days and 30% of animals survived beyond 40 days.

4.4. Olaparib

Olaparib, a PARP1 inhibitor that blocks PAR synthesis and impairs DNA repair, was studied by Hart et al. as a radiosensitizer in DMG [14]. They investigated whether FUS-mediated BBB opening could improve olaparib delivery to the pons and enhance its therapeutic effects in combination with radiotherapy. In vitro, patient-derived DMG cell lines (HSJD-DIPG-007 and HSJD-DIPG-011) showed that olaparib suppressed PAR synthesis, enhanced radiosensitivity, and delayed tumor regrowth, with prolonged low-dose exposure proving more effective than short high-dose exposure. In vivo, FUS with microbubbles increased pontine olaparib concentration 5.36-fold. In an orthotopic HSJD-DIPG-007 xenograft model, treatment arms including radiotherapy reduced primary tumor growth, and the combination of FUS, olaparib, and radiation therapy produced the strongest local suppression. However, overall survival was not improved due to rapid tumor dissemination and metastatic spread.

4.5. Dordaviprone

Dordaviprone (also known as ONC201) is a mitochondrial ClpP (caseinolytic protease P) activator that induces mitochondrial dysfunction and apoptosis through disruption of proteostasis and increased production of reactive oxygen species (ROS) [9]. It recently received FDA accelerated approval for progressive DMGs after positive responses in clinical trials.

Woldegerima et al. investigated whether FUS could enhance Dordaviprone delivery by transiently disrupting the BBB in a syngeneic diffuse midline glioma (DMG) mouse model [15]. BBB opening was confirmed with contrast-enhanced T1-weighted MRI, and treatment effects were validated though molecular and imaging analyses. In mice treated with ONC201+FUS, Western blots showed significantly decreased levels of NADH-ubiquinone oxidoreductase 1 alpha subcomplex 12 (NDUFA12), a pharmacodynamic biomarker of ONC201 response, and elevated ROS compared to mice treated with ONC201 alone. Mice in the combination group also demonstrated reduced tumor burden on imaging compared to those receiving ONC201 monotherapy, supporting the utility of FUS in augmenting ONC201 efficacy.

4.6. Radiation Therapy

Radiation therapy (RT) has been a mainstay of neuro-oncology in general, and its combination with FUS to potentiate therapeutic outcomes has been an area of great interest. Recent studies over the last two years have provided additional insight into the possible clinical benefits of this combination.

Chen et al. used a mouse glioblastoma model and compared three groups: RT along (2 Gy), RT and FUS (2 Gy), and RT (5 Gy) [16]. FUS was delivered immediately before radiation, targeting tumor regions. The study found that mice receiving FUS prior to RT (2 Gy) exhibited significantly longer survival compared to those receiving RT (2 Gy) alone or no treatment, though survival was comparable to the RT (5 Gy) group.

Fletcher et al. used both healthy rats and rats bearing F98 glioma tumors [17]. In healthy rats, the combination of FUS and RT at 8 and 15 Gy induced ablative lesions detectable by MRI within 72 h, persisting up to 21 days. In the F98 glioma model, FUS combined with RT (4 Gy), FUS alone, and combination therapy were compared. The FUS-RT group showed tumor volume reductions of 45–57% compared to controls. However, survival benefits were minimal. Histological analysis showed significant increases in apoptosis and vessel-associated ceramin in the FUS-RT group compared to FUS or RT alone.

Tazhibi et al. looked at the feasibility of brainstem FUS in combination with clinical doses of RT through a murine model of DMG [18]. The brainstems of male B6 (Cg)-Tyrc-2J/J albino mice were intracranially injected with mouse DMG cells (PDGFB^{+}, H3.3K27M, p53$^{-/-}$). A clinical RT dose of 39 Gy in 13 fractions (39 Gy/13fx) was delivered using the Small Animal Radiation Research Platform (SARRP) or XRAD-320 irradiator. FUS was administered via a 0.5 MHz transducer, with BBBO and tumor volume monitored by magnetic resonance imaging (MRI). Results showed good tolerance of FUS-mediated BBBO in combination with RT, with no impact on cardiorespiratory rate, motor function, or tissue integrity. Local control was obtained, though disease progression occurred after 3–4 weeks.

Gallitto et al. explored the use of napabucasin, an NAD(P)H quinone dehydrogenase 1 (NQO1)-bioactivatable reactive oxygen species (ROS)-inducer, as a potential therapeutic radiosensitizer in DMG18 [19]. Through the use of patient-derived DMG cultures, focused ultrasound (FUS) and convection-enhanced delivery (CED) were used to overcome the BBB and optimize targeted drug delivery. In combination with radiotherapy, this improved local control.

4.7. Immunotherapy

Monoclonal Anti-CD47 Antibody

CD47 is an immune checkpoint molecule overexpressed on tumor cells that interacts with singles on macrophages to inhibit phagocytosis. Anti-CD45 monoclonal antibodies block this interaction, enhancing macrophage-mediated clearance of cancer. Sheybani et al. demonstrated that timing of FUS delivery was critical for maximizing antibody uptake [20]. They used Zirconium-89-labeled anti-CD47 monoclonal antibodies in a glioma model and utilized PET imaging and ex vivo biodistribution to show that post-FUS antibody delivery led to significantly enhanced antibody accumulation in gliomas, compared to pre-FUS or controls. Treated mice showed slowed tumor progression and longer survival with reduced antibody dosage, indicating more efficient and targeted delivery.

4.8. Cetuximab

Cetuximab is a monoclonal antibody that inhibits cell proliferation and angiogenesis by targeting the epidermal growth factor receptor (EGFR) frequently found on tumor cells. Porret et al. tested cetuximab, an EGFR-targeting monoclonal antibody, using orthotopic U251 glioblastoma xenografts in nude mice [21]. FUS was applied at peak plasma levels to maximize delivery. FU significantly improved early delivery and homogenization of cetuximab in the brain, including tumor regions. However, long-term accumulation or retention of the antibody in tumors was not statistically different, and there was no significant difference in survival between FUS-treated and control groups.

4.9. Immune Checkpoint Inhibitors

Immune checkpoint inhibitors, such as Anti-PD1 and anti-CTLA4 antibodies, are commonly utilized in solid tumor treatments to reactivate the T-cell response against tumors. Anti-PD1 antibodies block PD-1 receptors, restoring cytotoxic activity. Anti-CTLA-1 antibodies enhance T-cell activation by blocking CTLA-4, which normally suppresses T-cell activation. However, like many other current therapeutics, they are limited by the BBB for use in brain tumors. Lee et al. tried these immunotherapies with FUS in orthotopic GL261 glioma bearing C57BL/6 mice [22]. They applied a closed-loop FUS system to precisely open the BBB before administering anti-PD1 and anti-CTLA4 antibodies. Flow cytometry and immunohistochemistry revealed increased CD8+ T-cell infiltration, reduced tumor growth, and improved survival compared to immune checkpoint blockade alone.

4.10. Biological Modifiers

Gambogic Acid

Gambogic acid is a natural compound that induces apoptosis by activating caspases and disrupting mitochondrial function. Dong et al. explored the use of gambogic acid-loaded poly(lactic-co-glycolic acid) (PLGA) nanoparticles for glioma therapy in U87 and U251 glioma-bearing mice [23]. FUS was applied at defined intervals post-injection to target tumor zones, and uptake was tracked using fluorescent dye-labeled particles. FUS treatment enhanced nanoparticle delivery to tumor cells by over 3-fold, and histological analysis confirmed increased apoptosis and tumor inhibition. FUS-treated mice had significantly smaller tumors and improved survival compared to controls.

4.11. Commentary

Across drug classes—including chemotherapies, radiosensitizers, immunotherapies, and biologics (Table 1)—common patterns emerge in terms of drug uptake, tumor control, and survival outcomes, while also highlighting specific challenges that warrant further investigation.

Table 1. Preclinical Focused Ultrasound Studies for DMG/DIPG.

Reference	Treatment	Dose	Animal model	Ultrasound Intensity or Frequency	BBB Permeability Measure	Results	Median Survival	Adverse Events
Kim et al. 2021 [10]	Doxorubicin	8 mg/kg	GL261 glioma-bearing mice and F98 glioma-bearing rats	1.7 MHz, 10 min at 41.5 °C	T2-weighted MR images	FUS-induced hyperthermia significantly enhanced the delivery of thermosensitive liposomal doxorubicin to tumore, improved vascular permeability, and prolonged survival of glioma-bearing rodents.	16 days (vs. 9 days in the TSL-Dox alone)	Hemorrhage in the brain near the skull surface when hyperthermia was applied at 42.5 °C for 10 min in mice. Lower hyperthermia (41.5 °C) was safe in rodents, with no damage seen.
Choi et al., 2025 [11]	Doxorubicin	5.67 mg/kg	Male Sprague Dawley rats implanted with intracranial 9L gliosarcoma tumors	1 MHz, 120 s duration, acoustic pressure of 0.5, 1, 2, or 0.72 Mpa	T1-weighted or Dynamic contrast-enhanced MRI, Evans blue dye staining	FUS stimulation without microbubbles followed by BBB disruption with microbubbles significantly increased blood-tumor permeability and doxorubicin delivery compared to BBB disruption alone. Dynamic contrast-enhanced MRI showed a 2.65-fold increase in signal intensity and a 2.08-fold increase in permeability in tumor regions.	N/A	None detected
Yang et al. 2021 [12]	Temozolomide (TMZ)	50 mg/kg daily for 5 days	Orthotopic glioblastoma-bearing nude mice implanted with human T98G glioblastoma cells	1.84 W, 3–5 m duration	Evans blue dye staining	FUS improved accumulation of lipid-polymer hybrid nanoparticles (LPHNs) in tumor regions and enabled CRISPR/Cas9-mediated MGMT knockdown, restoring sensistivity to TMZ. This combination therapy inhibited tumor growth and significantly extended survival in glioma-bearing mice.	43 days (vs. 22–30 days in other groups)	At higher FUS/MB doses: erythrocyte extravasation and hemorrhage during BBB opening.
Shen et al., 2022 [13]	Epirubicin	2 mg/kg	Nude mice baring intracranial glioma xenografts	0.971 MHz, 60 s duration, 0.52 MPa	In vivo fluorescence imaging using an IVIS imaging system	FUS treatment significantly increased epirubicin accumulation in glioma tissue (4.4-fold enhancement over controls). In vivo imaging and histology confirmed deep tumore penetration and prolonged survival (50% increase in median survival).	36.5 days (vs. 25 days in other groups)	None detected

Table 1. *Cont.*

Reference	Treatment	Dose	Animal model	Ultrasound Intensity or Frequency	BBB Permeability Measure	Results	Median Survival	Adverse Events
Hart et al., 2023 [14]	Olaparib	10 or 100 mg/kg	Patient-derived xenograft (PDX) DMG mouse model.	1 MHz, 1.6 Hz pulse repetition frequency, 400 kPa pressure, 120 s duration	Evans blue dye staining	Effects of PARP1 inhibition were evaluated in vitro using viability, clonogenic, and neurosphere assays. In vivo olaparib extravasation and pharmacokinetic profiling following FUS-BBBO was measured by LC-MS/MS. Survival benefit of FUS-BBBO combined with olaparib and RT was assessed.	N/A	None detected
Woldegerima et al., 2024 [15]	Dordaviprone (ONC201)	N/A	Syngeneic diffuse midline glioma mouse	N/A	Gadolinium-enhanced MRI	FUS enhanced ONC201 delivery, leading to greater biomarker response, increased ROS, and reduced tumor burden compared to ONC201 alone.	N/A	None detected
Chen et al., 2023 [16]	Radiotherapy (RT)	Whole brain, 2 and 5 Gy at a rate of 3.3 Gy/min	Mouse glioblastoma (GBM) model	500 kHz, 0.4–0.56 mechanical index (MI), duty cycle of 1%, burst period of 1 s, 120 s duration	MRI	Mice recieving FUS prior to RT (2 Gy) exhibited significantly longer survival compared to those recieving RT (2 Gy) alone or no treatment, though survival was comparable to the RT (5 Gy) group.	N/A	None detected across 24 RT-FUS sessions; one grade-3 radiation necrosis attributed to re-irradiation (RT)
Fletcher et al., 2024 [17]	Radiotherapy (RT)	4, 8, 15 Gy	Healthy rats and rats bearing F98 glioma tumors	220 kHz, 102–444 kPa	Contrast enhanced T1-weighted MRI	In healthy rats, the combination of FUS and RT at 8 and 15 Gy induced ablative lesions detectable by MRI within 72 h, persisting up to 21 days. In the F98 glioma model, FUS combined with 4 Gy RT reduced tumore volumes by 45–57% compared to controls. However, survival benefits were minimal. Histological analysis showed significant increases in apoptosis and vessel-associated ceramin in the FUS-RT group compared to FUS or RT alone.	28 days (vs. 27 days in control and RT only groups)	Transient edema on MRI; rare minor T2 change (1/13 rats); no motor/neurologic deficits. FUS+RT at high dose: MRI-visible lesions and histologic scarring; authors caution on MB dose/pressure escalation.

Table 1. *Cont.*

Reference	Treatment	Dose	Animal model	Ultrasound Intensity or Frequency	BBB Permeability Measure	Results	Median Survival	Adverse Events
Tazhibi et al., 2024 [18]	Radiotherapy (RT)	39 Gy in 13 fractions	Non–tumor-bearing mice and syngeneic DMG murine model	0.5 MHz, peak-negative pressure 0.3 Mpa, 5 Hz repetition time, 120 s duration	MRI	Demonstrated that repeated brainstem FUS during RT is safe, feasible, and well-tolerated; progression still occurred post-RT	54 days (vs. 28 days in control)	None detected
Gallitto et al., 2025 [19]	Napabucasin + Radiotherapy (RT)	2 Gy in 5 fractions; Napabucasin 80 µM	Patient-derived DMG cultures; orthotopic DMG mouse	N/A	Gadolinium contrast enhancement on MRI	Napabucasin acted as a potent radiosensitizer; CED delivery improved survival in vivo.	46 days (vs. 33 days RT alone and 26 days napabucasin alone)	None detected
Sheybani et al. 2021 [20]	Monoclonal anti-CD47 antibody (mcD47)	8 or 32 mg/kg every 3 days for 3 doses	Orthotopic murine glioma model using GL261 cells implanted in C57BL/6 mice	1.1 MHz, 0.5% duty cycle, 0.4 MPa, 2 m duration	Contrast-enhanced MR imaging	Post-FUS delivery of Zirconium-89 labeled mCD47 led to significantly enhanced antibody accumulation in gliomas, compared to pre-FUS administration. This treatment sequence suppressed tumor growth and prolonged survival using less antibody than prior methods.	Increased 14 day survival by 40%	None detected
Porret et al., 2023 [21]	Cetuximab	N/A	Orthotopic U251 glioblastoma xenografts in nude mice	1 MHz, 0.5% duty cycle, 60 s duration	Evans blue dye staining	FUS significantly increased early deliver and homogenization of cetuximab in the brain, including tumor regions. However, it did not enhance long-term accumulation or retention of the antibody in tumors. There was no significant difference in survival between FUS-treated and control groups.	N/A	None detected

Table 1. *Cont.*

Reference	Treatment	Dose	Animal model	Ultrasound Intensity or Frequency	BBB Permeability Measure	Results	Median Survival	Adverse Events
Lee et al., 2022 [22]	Immune checkpoint inhibitors (anti-PD-1 and anti-CTLA-4 antibodies	100 mg/kg	Orthotopic GL261 glioma bearing C57BL/6 mice	1.64 MHz, 10-ms pulse length, 1-Hz pulse repetition frequency, 120 s duration, 70-kPa peak negative pressure	Dynamic contrast-enhanced MRI	Closed-loop controlled FUS precisely opened the BBB, enhancing delivery of immune checkpoint inhibitors to brain tumores. This combination therapy increased infiltration of cyctotoxic T cells, reduced tumor growth, and improved survival compared to immune checkpoint blockade alone.	N/A	Minimal inflammatory effects outside tumor; one animal death due to anesthesia (not FUS-related)
Dong et al. 2022 [23]	Gambogic Acid	50 µL (1.5 µmol/L)	U87 and U251 glioma-bearing mice	900 W, 0.6% duty cycle, 2 m duration, 0.63 Mpa	Fluorescence signal was detected with a live animal imaging system (IVIS)	FUS enhanced delivery of Gamogic acid-loaded poly(lactic-co-glycolic acid) (PLGA) nanoparticles to glioma cells, increasing drug uptake and tumor inhibition.	59 days (vs. 38-56 days in other groups)	Higher intensities (0.76–0.88 MPa) caused erythrocyte extravasation and apoptosis in adjacent tissue

Among chemotherapeutics, anthracyclines such as doxorubicin and epirubicin—which typically fail to penetrate the BBB—demonstrated significantly enhanced intratumoral accumulation when paired with FUS. In Kim et al.'s study, thermosensitive liposomal doxorubicin (TSL-Dox) combined with FUS hyperthermia achieved 3.5- to 5-fold increases in drug uptake and prolonged survival in glioma-bearing rodents [10]. Similarly, Shen et al. reported a 4.4-fold increase in epirubicin delivery with DNA nanocarriers and microbubble-assisted FUS, resulting in nearly 87% tumor growth inhibition and extended survival [13]. These effects are not limited to traditional cytotoxics. Yang et al. combined FUS with a CRISPR/Cas9-loaded nanoparticle targeting MGMT, a resistance mediator in temozolomide (TMZ) therapy [12]. This approach sensitized glioblastoma cells to TMZ and doubled median survival compared to TMZ monotherapy. Likewise, ONC201 delivery was markedly improved by FUS in a syngeneic DMG model, leading to enhanced ROS generation and tumor reduction [15]. Collectively, these studies demonstrate that FUS not only facilitates delivery of impermeable drugs, but also enables combinatorial strategies—such as gene editing or thermosensitive release—unfeasible under standard conditions.

The studies combining low-intensity focused ultrasound (LIFU) with radiotherapy (RT) have shown promise in improving local tumor control, but survival benefits remain limited—highlighting a critical translational gap. While FUS evidently enhances local effects of RT—via improved drug delivery, BBB permeability, and localized apoptosis—the limited systemic impact, presence of infiltrative disease, and short follow-up periods may explain the modest survival outcomes. Although LIFU-enhanced RT may achieve precise, localized tumor ablation or radiosensitization, these effects do not address micrometastatic disease or diffuse infiltrative tumor cells, which are characteristic of high-grade gliomas. Consequently, local tumor regression may not translate into meaningful survival gains if systemic or distant tumor burden remains unaddressed. Moreover, the timing, dose, and fractionation of radiotherapy relative to LIFU exposure may significantly influence outcomes. For instance, suboptimal synchronization between BBB opening and drug/radiation delivery could reduce therapeutic efficacy. In some studies, BBB disruption may have occurred outside the window of maximal radiosensitization or immune activation, limiting synergistic effects. Additionally, the total radiation dose used in certain models may have been insufficient to produce durable responses, especially when not paired with effective systemic agents. A more durable effect may require integrating FUS-enhanced RT with systemic or immunomodulatory therapies in future studies.

Immunotherapy, while transformative in other cancers, has underperformed in brain tumors due to limited immune infiltration and poor BBB penetration of monoclonal antibodies. FUS appears to directly address both issues. In glioma models, Sheybani et al. demonstrated that post-FUS delivery of radiolabeled anti-CD47 antibodies significantly improved antibody accumulation and survival, even at reduced doses [20]. Similarly, Lee et al. showed that FUS-enhanced delivery of anti-PD1 and anti-CTLA4 antibodies led to increased CD8+ T-cell infiltration, reduced tumor burden, and prolonged survival compared to checkpoint inhibitors alone [22].

Overall, focused ultrasound (FUS) in combination with therapeutic agents or radiotherapy resulted in observable BBB opening in 6 out of 7 studies, confirmed by imaging modalities such as gadolinium-enhanced MRI or histological tracers. Reported signal enhancement or tracer penetration typically ranged from approximately 20% to over 50%, indicating consistent, though variable, permeability effects depending on parameters and delivery method. In terms of therapeutic efficacy, survival benefit ranged from negligible to significant, with increases of up to ~50% in median survival observed in models treated with temozolomide, doxorubicin, epirubicin, ONC201, and radiotherapy. In contrast, cetuximab administration showed enhanced delivery without a corresponding survival improvement. Importantly, adverse effects were minimal in most studies, with no detectable neurotoxicity reported in 5 of 7 investigations. The only observed complications included mild thermal damage at higher hyperthermia thresholds (≥42.5 °C) and radiation necrosis in re-irradiation contexts, not directly attributable to FUS. These composite findings support the safety and potential efficacy of FUS-mediated delivery in glioma models and underscore the need for standardized reporting of key outcome measures to facilitate future meta-analyses.

Important to note is that when extrapolating data from adult glioblastoma models to pediatric diffuse midline gliomas (DMG), several critical differences must be considered. Pediatric DMGs, such as those harboring the H3K27M mutation, have distinct molecular profiles, epigenetic landscapes, and tumor microenvironments compared to adult GBMs. These tumors often arise in anatomically and functionally sensitive regions like the brainstem, where therapeutic windows are narrower and safety margins are limited. Moreover, the developing pediatric brain differs significantly in vascular permeability, immune response, and tissue composition, all of which can influence how focused ultrasound interacts with both tumor and surrounding tissue. The pharmacokinetics of therapeutics delivered via blood–brain barrier disruption may also differ in children, impacting efficacy and safety. As such, while adult GBM models offer valuable proof-of-concept data, caution is warranted when interpreting these results in the pediatric context. There remains an urgent need for preclinical models that faithfully replicate the biological and anatomical characteristics of pediatric DMG to guide the safe and effective translation of focused ultrasound technologies.

5. Clinical Trials Involving Focused Ultrasound for Pediatric DMG

As FUS has exhibited promising safety and efficacy in preclinical studies, extension into the clinical field has continued to expand over recent years.

Translating transcranial applications of FUS from pre-clinical models such as rodents to humans requires multiple parameter optimization steps. Often, non-human primate models are employed for this parameter optimization due the similarities in skull thickness and brain size, however significant parameter optimization is also required during clinical trials. The main parameter that needs to be changed from pre-clinical to clinical applications is the ultrasound frequency, as a lower frequency is required for efficient transcranial FUS insertion without significant energy loss due to attenuation. FUS applications in human clinical trials use frequencies such as 0.25 MHz [24], which are significantly lower than pre-clinical models. Other parameters such as power and time are still being optimized in pre-clinical models and can be leveraged to affect the resulting BBB opening and drug/gene delivery.

Experience from adult LIFU protocols, both in malignant and non-malignant conditions, provides a valuable reference point. Adult studies have established dosing thresholds, thermal safety limits, and targeting strategies under image guidance, which can inform the development of pediatric protocols. Early clinical experiences—though limited—suggest the feasibility of LIFU in selected pediatric cases when protocols are appropriately modified. Drawing on adult data while accounting for pediatric-specific anatomical and developmental considerations can provide a pragmatic and evidence-based roadmap for expanding LIFU applications in younger populations, as is currently being seen with four ongoing LIFU clinical trials in pediatrics (Table 2).

Table 2. Focused Ultrasound for Pediatric DMG/DIPG clinical trials.

Title	NCT	Status	Location	Size	Device	Drug
Non-Invasive Focused Ultrasound (FUS) With Oral Panobinostat in Children With Progressive Diffuse Midline Glioma (DMG)	NCT04804709	Active, not recruiting	New York, NY	3	Focused Ultrasound with neuro-navigator-controlled sonication	Panobinostat
A Feasibility Study Examining the Use of Non-Invasive Focused Ultrasound (FUS) With Oral Etoposide Administration in Children With Progressive Diffuse Midline Glioma (DMG)	NCT05762419	Recruiting	New York, NY	10	Focused ultrasound with neuro-navigator-controlled sonication	Etoposide
Blood Brain Barrier (BBB) Disruption Using Exablate Focused Ultrasound With Doxorubicin for Treatment of Pediatric Diffuse Intrinsic Pontine Gliomas (DIPG)	NCT05630209	Recruiting	Washington, DC and Miami, FL	10	Exablate	Doxorubicin
Blood Brain Barrier (BBB) Disruption Using Exablate Focused Ultrasound With Doxorubicin for Treatment of Pediatric DIPG	NCT05615623	Recruiting	Toronto, ON	10	Exablate	Doxorubicin

The first ever clinical trial of FUS for pediatric DIPG/DMG at Columbia University (NCT04804709) started in 2020 and used the Delsona device to open the BBB in conjunction with oral Panobinostat [25]. As a phase 1 trial, the primary outcome was safety, with secondary outcome measures looking at 6-month progression free survival and blood–brain barrier/tumor imaging changes. Eligible participants were aged 4 to 21 with confirmed pontine or thalamic DMG, able to tolerate anesthesia, and able to swallow capsules. Key exclusions included active infections, coagulopathies, prior cerebrovascular events, or contraindications to panobinostat or MRI contrast. The trial enrolled five patients (three accrued), and FUS successfully achieved BBB opening in most procedures. Only mild, transient side effects were reported, with one case of Grade 1 dermatologic toxicity. However, the study was terminated early due to the withdrawal of panobinostat from the market. Despite early closure, the findings demonstrated that BBB disruption with FUS was feasible and well tolerated. The trial is now re-opened and recruiting, with the use of oral etoposide as replacement (NCT05762419) [26].

The other two ongoing clinical trials (NCT05630209 and NCT05615623) utilize Insightec's Exablate Model 4000 to disrupt the blood–brain barrier [27]. All three institutions involved- Children's National Hospital, Nicklaus Children's Hospital, and Sunnybrook Research Institute, are also using doxorubicin as the chemotherapy delivered concurrently to the tumor site during the temporary BBB opening. As Phase 1/2 studies, the goal is to evaluate the safety, feasibility, and preliminary efficacy of using magnetic resonance–guided focused ultrasound (MRgFUS) to transiently open the blood–brain barrier (BBB) and enhance the delivery of doxorubicin in pediatric patients with diffuse intrinsic pontine glioma (DIPG). Both studies, which are currently recruiting, include children aged 5 to 18 years who have completed radiation therapy and have a life expectancy of at least six months. Key inclusion criteria also require stable steroid use and confirmed DIPG diagnosis, while exclusion criteria include prior maximum anthracycline exposure, bleeding disorders, and contraindications to doxorubicin, ultrasound contrast agents, or MRI. The primary endpoints are the safety and feasibility of BBB opening, with secondary endpoints focusing on imaging outcomes and survival metrics. Results have not yet been reported as the trials are ongoing.

6. Future Research

With continued advancements in both preclinical and clinical focused ultrasound research, several promising avenues for future investigation in the treatment of pediatric diffuse midline gliomas. Preclinical studies have shown that low-intensity focused ultrasound can transiently disrupt the blood–brain barrier, enhancing the delivery and bioavailability of chemotherapeutic agents like etoposide and doxorubicin that otherwise exhibit poor CNS penetration [28,29]. This represents a critical opportunity, as systemic therapy for DMG has historically been limited by the inability to achieve therapeutic drug concentrations within the tumor [30].

Expanding the repertoire of systemically administered agents tested in combination with FUS is an important next step. This includes not only traditional chemotherapies but also targeted drugs and new agents being developed [31]. A rational approach to drug selection will be essential, with emphasis on agents whose pharmacokinetics are compatible with the temporal window of BBB permeability following FUS. Integration of radiolabeling or other molecular imaging techniques to confirm increased intratumoral concentrations in real time will be vital for validating the efficacy of delivery and for guiding dose optimization in early-phase clinical trials.

In parallel, the combination of FUS with other treatment modalities offers an exciting area of exploration. Radiotherapy remains the standard of care in DMG, and FUS may enhance the radiosensitivity of tumor tissue through mechanisms such as vascular disruption or modulation of DNA repair pathways. Moreover, increasing attention is being paid to the potential immunomodulatory effects of FUS. Studies have shown that FUS-mediated BBB opening leads to increased infiltration of immune cells, including microglia and macrophages, into the CNS in both healthy and tumor-bearing brains [30,32]. This important role of the immune system on patterns of progression in malignancy could be leveraged by combination approaches with immune checkpoint inhibitors, oncolytic viruses, or adoptive cell therapies—therapies that have previously been limited by the immunologically "cold" nature of DMG [33].

Future research should prioritize in-depth characterization of the immune microenvironment in DMG before and after FUS, using single-cell sequencing, spatial transcriptomics, and immunohistochemistry to elucidate dynamic changes in immune cell composition and function [31,34]. These studies may uncover biomarkers predictive of treatment response and inform the timing and sequencing of immunotherapy relative to FUS application.

To advance the translational application of LIFU in pediatrics as FUS platforms and targeting precision continue to improve, it is essential to consider both regulatory and anatomical dimensions with greater specificity [35]. From a regulatory standpoint, the U.S. FDA offers several pathways that may facilitate the clinical translation of LIFU technologies in children. The Humanitarian Device Exemption (HDE) allows for the approval of devices intended to treat rare conditions (affecting fewer than 8000 individuals per year in the U.S.), which may be particularly relevant given the low incidence of some pediatric neurological disorders. Additionally, the Early Feasibility Study (EFS) program supports initial clinical evaluations of novel devices in a small number of patients, allowing for protocol refinements based on early findings. These frameworks provide flexible mechanisms for generating safety and feasibility data in pediatric populations, which could be used to support expanded indications in the future.

Anatomically, children present unique challenges for LIFU application. Important considerations include the heterogeneity of gray and white matter, distribution of vasculature, and thickness and unevenness of skulls (skull thickness is generally lower in younger children, especially infants) which must be considerations when designing FUS devices as this can significantly affect acoustic transmission and focal accuracy [36]. Historically, transcranial FUS has been limited by variations in skull thickness that result in distortion of the ultrasound beams, but developments of hemispherical phased-array transducers and software to correct for phase aberrations have been overcoming these challenges since the 1990s and continue to progress [6]. The presence of open fontanelles in infants also offers a potential acoustic window, which may facilitate more precise targeting without skull interference, but also raises concerns about vulnerability to thermal or mechanical damage. Brain maturation further complicates treatment planning, as the developmental characteristics of children's brain tissue—such as higher water content and ongoing myelination—pose unique challenges for energy delivery and safety. Current technical bottlenecks include precise real-time targeting through small or irregular acoustic windows, managing heat deposition in delicate tissues, and developing age-specific models for treatment planning. Some limitations, such as achieving consistent energy penetration across a wide range of pediatric skull anatomies, remain unresolved and require further innovation in transducer design and adaptive focusing algorithms. These factors underscore the need for age-specific calibration of dosing parameters, targeting algorithms, and safety margins. Future clinical trials will need to address these challenges.

Finally, both immediate and long-term monitoring for potential off-target effects, especially in the sensitive structures of the brainstem, is vital [37]. Real-time FUS treatment monitoring is crucial for ensuring safety and for advancing clinical adoption. In MR-guided FUS, thermal monitoring is possible with MR thermometry, which would help avoid tissue overheating and thermal damaging effects. However, with clinical systems that are located outside the MRI, this type of monitoring is not possible. Acoustic monitoring of microbubble activity (cavitation) is therefore an essential component of US-guided FUS clinical systems, as it enables the real-time assessment of microbubble activity and the avoidance of damaging violent microbubble collapse, known as inertial cavitation. However, cavitation monitoring still requires additional development for reliable transcranial microbubble monitoring, and further research on the cavitation thresholds for safe and effective BBB opening needs to take place before widespread clinical adoption. In parallel, longitudinal assessments of neurocognitive function and quality of life will be essential to ensure that FUS-based interventions do not compromise neurologic integrity, particularly in a developing pediatric brain.

Overall, the integration of FUS into multi-modal treatment strategies for pediatric DMG holds substantial promise. The next generation of preclinical models and clinical trials must be thoughtfully designed to leverage this technology's full potential while maintaining a rigorous focus on safety, efficacy, and translational feasibility.

7. Conclusions

Diffuse midline gliomas, including DIPG, continue to present an urgent clinical challenge in the realm of pediatric oncology, with limited effective treatment options and dismal prognoses. While focused ultrasound is a technology that has existed since the 1940s, recent advances represent a paradigm shift in the approach to drug delivery for these inoperable brainstem tumors, offering a noninvasive method to transiently disrupt the blood–brain barrier and enhance therapeutic efficacy [38,39]. Preclinical studies have consistently demonstrated the potential of LIFU to increase intratumoral delivery of chemotherapeutics, with recent advances in different medications being utilized. Further clinical trial openings will help to elucidate further on this approach, which is both feasible and safe in pediatric populations, but long-term outcomes and survival benefits remain unknown.

Continued interdisciplinary collaboration and well-designed clinical trials will be essential to fully translate the promise of LIFU into a viable adjunct for treating DMGs in the pediatric population, as most trials are currently focused on adults [40,41]. As the field evolves, focused ultrasound may emerge as a cornerstone technology in overcoming the longstanding therapeutic barriers of pediatric neuro-oncology [2,42]. Moreover, integrating LIFU with targeted therapies or immunotherapeutics may further amplify its clinical benefits, particularly in a precision medicine framework tailored to the molecular profiles of individual tumors.

Nonetheless, several challenges must be addressed before LIFU can become a routine clinical modality. These include optimizing sonication parameters for maximal drug delivery with minimal off-target effects, developing robust imaging and biomarker tools to monitor treatment response, and ensuring equitable access to this technology across diverse healthcare settings. Understanding the biological response of tumor and normal brain tissue to repeated BBB disruption also remains a critical area of ongoing investigation, as a significant limitation remains that preclinical data has thus far been isolated to specific cell lines and are not yet validated in multiple models.

In addition to technical and biological considerations, the feasibility of widespread clinical adoption of low-intensity focused ultrasound (LIFU) is also influenced by economic and logistical factors. Currently, the cost of LIFU systems remains high due to the complexity of the technology and limited market availability. This includes not only the equipment itself but also the infrastructure required for real-time monitoring with MRI. Furthermore, specialized training is essential for clinicians and technical staff to ensure safe and effective operation, which adds to the initial implementation burden. Compared to traditional treatments such as surgery or radiotherapy, which are already integrated into hospital workflows, LIFU may face higher upfront costs, though its noninvasive nature could reduce long-term expenses related to hospital stays, complications, and rehabilitation. To improve cost-effectiveness, strategies such as shared device models across institutions, development of simplified user interfaces, and standardized training programs could help reduce barriers to entry. Future health economic analyses are needed to quantitatively compare the cost–benefit profiles of LIFU and conventional therapies, particularly in pediatric populations where long-term outcomes and quality of life are especially critical.

The potential applications of LIFU extend beyond DMGs, suggesting a broader relevance for other central nervous system malignancies and neurodegenerative disorders where BBB permeability limits therapeutic efficacy. In this context, the current momentum surrounding LIFU research signals a transformative moment in neuro-oncology, one that holds promise not only for extending survival but also for improving the quality of life in a population of patients who have long faced limited hope.

Funding: This research received no external funding.

Conflicts of Interest: The authors declare no conflict of interest.

References

1. Chesney, K.M.; Keating, G.F.; Patel, N.; Kilburn, L.; Fonseca, A.; Wu, C.C.; Nazarian, J.; Packer, R.J.; Donoho, D.A.; Oluigbo, C.; et al. The role of focused ultrasound for pediatric brain tumors: Current insights and future implications on treatment strategies. *Childs Nerv. Syst.* **2024**, *40*, 2333–2344. [CrossRef]
2. Udaka, Y.T.; Packer, R.J. Pediatric Brain Tumors. *Neurol. Clin.* **2018**, *36*, 533–556. [CrossRef]
3. Ostrom, Q.T.; Price, M.; Ryan, K.; Edelson, J.; Neff, C.; Cioffi, G.; Waite, K.A.; Kruchko, C.; Barnholtz-Sloan, J.S. CBTRUS Statistical Report: Pediatric Brain Tumor Foundation Childhood and Adolescent Primary Brain and Other Central Nervous System Tumors Diagnosed in the United States in 2014–2018. *Neuro-Oncology* **2022**, *24* (Suppl. 3), iii1–iii38. [CrossRef]

4. Parekh, K.; LeBlang, S.; Nazarian, J.; Mueller, S.; Zacharoulis, S.; Hynynen, K.; Powlovich, L. Past, present and future of Focused Ultrasound as an adjunct or complement to DIPG/DMG therapy: A consensus of the 2021 FUSF DIPG meeting. *Neoplasia* **2023**, *37*, 100876. [CrossRef]
5. Janwadkar, R.; Leblang, S.; Ghanouni, P.; Brenner, J.; Ragheb, J.; Hennekens, C.H.; Kim, A.; Sharma, K. Focused Ultrasound for Pediatric Diseases. *Pediatrics* **2022**, *149*, e2021052714. [CrossRef]
6. Hersh, A.M.; Jillala, R.; Kramer, P.; Keating, R.F.; Syed, H.R.; Oluigbo, C.O.; Groves, M.L.; Manbachi, A.; Theodore, N.; Woodworth, G.F.; et al. Focused ultrasound in pediatric neurosurgery: A scoping review of opportunities and challenges. *Childs Nerv. Syst.* **2025**, *41*, 210. [CrossRef] [PubMed]
7. Escoffre, J.M.; Piron, J.; Novell, A.; Bouakaz, A. Doxorubicin delivery into tumor cells with ultrasound and microbubbles. *Mol. Pharm.* **2011**, *8*, 799–806. [CrossRef]
8. Sewing, A.C.; Lagerweij, T.; Van Vuurden, D.; Meel, M.; Veringa, S.; Carcaboso, A.; Gaillard, P.; Vandertop, W.; Wesseling, P.; Noske, D.; et al. Preclinical evaluation of convection-enhanced delivery of liposomal doxorubicin to treat pediatric diffuse intrinsic pontine glioma and thalamic high-grade glioma. *J. Neurosurg. Pediatr.* **2017**, *19*, 518–530. [CrossRef] [PubMed]
9. Filieri, S.; Miciaccia, M.; Armenise, D.; Baldelli, O.M.; Liturri, A.; Ferorelli, S.; Sardanelli, A.M.; Perrone, M.G.; Scilimati, A. Can Focused Ultrasound Overcome the Failure of Chemotherapy in Treating Pediatric Diffuse Intrinsic Pontine Glioma Due to a Blood-Brain Barrier Obstacle? *Pharmaceuticals* **2025**, *18*, 525. [CrossRef] [PubMed]
10. Kim, C.; Guo, Y.; Velalopoulou, A.; Leisen, J.; Motamarry, A.; Ramajayam, K.; Aryal, M.; Haemmerich, D.; Arvanitis, C.D. Closed-loop trans-skull ultrasound hyperthermia leads to improved drug delivery from thermosensitive drugs and promotes changes in vascular transport dynamics in brain tumors. *Theranostics* **2021**, *11*, 7276–7293. [CrossRef]
11. Choi, H.J.; Han, M.; Jung, B.; Huh, H.; Lee, E.H.; Choi, J.R.; Park, J. Evaluation of blood-tumor barrier permeability and doxorubicin delivery in rat brain tumor models using additional focused ultrasound stimulation. *Sci. Rep.* **2025**, *15*, 6592. [CrossRef]
12. Yang, Q.; Zhou, Y.; Chen, J.; Huang, N.; Wang, Z.; Cheng, Y. Gene Therapy for Drug-Resistant Glioblastoma via Lipid-Polymer Hybrid Nanoparticles Combined with Focused Ultrasound. *Int. J. Nanomed.* **2021**, *16*, 185–199. [CrossRef]
13. Shen, Y.; Hu, M.; Li, W.; Chen, Y.; Xu, Y.; Sun, L.; Liu, D.; Chen, S.; Gu, Y.; Ma, Y.; et al. Delivery of DNA octahedra enhanced by focused ultrasound with microbubbles for glioma therapy. *J. Control. Release* **2022**, *350*, 158–174. [CrossRef]
14. Hart, E.t.; Bianco, J.; Bruin, M.A.C.; Derieppe, M.; Besse, H.C.; Berkhout, K.; Chin Joe Kie, L.A.; Su, Y.; Hoving, E.W.; Huitema, A.D.R.; et al. Radiosensitisation by olaparib through focused ultrasound delivery in a diffuse midline glioma model. *J. Control. Release* **2023**, *357*, 287–298. [CrossRef] [PubMed]
15. Woldegerima, A.; Wei, H.J.; Zhang, C.; Yadavillli, S.; Packer, R.; Wu, C.C.; Nazarian, J. DIPG-73. Focused Ultrasound for Treatment of Children Diagnosed with Diffuse Midline Gliomas. *Neuro-Oncology* **2024**, *26* (Suppl. 4). [CrossRef]
16. Chen, K.T.; Huang, C.Y.; Pai, P.C.; Yang, W.C.; Tseng, C.K.; Tsai, H.C.; Li, J.C.; Chuang, C.C.; Hsu, P.W.; Lee, C.C.; et al. Focused ultrasound combined with radiotherapy for malignant brain tumor: A preclinical and clinical study. *J. Neuro-Oncol.* **2023**, *165*, 535–545. [CrossRef]
17. Fletcher, S.P.; Chisholm, A.; Lavelle, M.; Guthier, R.; Zhang, Y.; Power, C.; Berbeco, R.; McDannold, N. A study combining microbubble-mediated focused ultrasound and radiation therapy in the healthy rat brain and a F98 glioma model. *Sci. Rep.* **2024**, *14*, 4831. [CrossRef]
18. Tazhibi, M.; McQuillan, N.; Wei, H.J.; Gallitto, M.; Bendau, E.; Webster Carrion, A.; Berg, X.; Kokossis, D.; Zhang, X.; Zhang, Z.; et al. Focused ultrasound-mediated blood-brain barrier opening is safe and feasible with moderately hypofractionated radiotherapy for brainstem diffuse midline glioma. *J. Transl. Med.* **2024**, *22*, 320. [CrossRef] [PubMed]
19. Gallitto, M.; Zhang, X.; De Los Santos, G.; Wei, H.J.; Fernández, E.C.; Duan, S.; Sedor, G.; Yoh, N.; Kokossis, D.; Angel, J.C.; et al. Targeted delivery of napabucasin with radiotherapy improves outcomes in diffuse midline glioma. *Neuro-Oncology* **2025**, *27*, 795–810. [CrossRef]
20. Sheybani, N.D.; Breza, V.R.; Paul, S.; McCauley, K.S.; Berr, S.S.; Miller, G.W.; Neumann, K.D.; Price, R.J. ImmunoPET-informed sequence for focused ultrasound-targeted mCD47 blockade controls glioma. *J. Control. Release* **2021**, *331*, 19–29. [CrossRef] [PubMed]
21. Porret, E.; Kereselidze, D.; Dauba, A.; Schweitzer-Chaput, A.; Jegot, B.; Selingue, E.; Tournier, N.; Larrat, B.; Novell, A.; Truillet, C. Refining the delivery and therapeutic efficacy of cetuximab using focused ultrasound in a mouse model of glioblastoma: An (89)Zr-cetuximab immunoPET study. *Eur. J. Pharm. Biopharm.* **2023**, *182*, 141–151. [CrossRef]
22. Lee, H.; Guo, Y.; Ross, J.L.; Schoen, S., Jr.; Degertekin, F.L.; Arvanitis, C. Spatially targeted brain cancer immunotherapy with closed-loop controlled focused ultrasound and immune checkpoint blockade. *Sci. Adv.* **2022**, *8*, eadd2288. [CrossRef]
23. Dong, L.; Li, N.; Wei, X.; Wang, Y.; Chang, L.; Wu, H.; Song, L.; Guo, K.; Chang, Y.; Yin, Y.; et al. A Gambogic Acid-Loaded Delivery System Mediated by Ultrasound-Targeted Microbubble Destruction: A Promising Therapy Method for Malignant Cerebral Glioma. *Int. J. Nanomed.* **2022**, *17*, 2001–2017. [CrossRef]

24. Bae, S.; Liu, K.; Pouliopoulos, A.N.; Ji, R.; Jiménez-Gambín, S.; Yousefian, O.; Kline-Schoder, A.R.; Batts, A.J.; Tsitsos, F.N.; Kokossis, D.; et al. Transcranial blood-brain barrier opening in Alzheimer's disease patients using a portable focused ultrasound system with real-time 2-D cavitation mapping. *Theranostics* **2024**, *14*, 4519–4535. [CrossRef]
25. Columbia University. Non-Invasive Focused Ultrasound (FUS) with Oral Panobinostat in Children with Progressive Diffuse Midline Glioma (DMG). Available online: https://clinicaltrials.gov/study/NCT04804709?cond=%22Diffuse%20Intrinsic%20Pontine%20Glioma%22&intr=%22Panobinostat%22&viewType=Table&rank=5 (accessed on 13 May 2025).
26. Columbia University. FUS Etoposide for DMG. Available online: https://www.clinicaltrials.gov/study/NCT05762419 (accessed on 13 May 2025).
27. InSightec. Blood Brain Barrier (BBB) Disruption Using Exablate Focused Ultrasound with Doxorubicin for Treatment of Pediatric Diffuse Intrinsic Pontine Gliomas (DIPG). Available online: https://clinicaltrials.gov/study/NCT05630209 (accessed on 13 May 2025).
28. Englander, Z.K.; Wei, H.J.; Pouliopoulos, A.N.; Bendau, E.; Upadhyayula, P.; Jan, C.I.; Spinazzi, E.F.; Yoh, N.; Tazhibi, M.; McQuillan, N.M.; et al. Focused ultrasound mediated blood-brain barrier opening is safe and feasible in a murine pontine glioma model. *Sci. Rep.* **2021**, *11*, 6521. [CrossRef] [PubMed]
29. Alli, S.; Figueiredo, C.A.; Golbourn, B.; Sabha, N.; Wu, M.Y.; Bondoc, A.; Luck, A.; Coluccia, D.; Maslink, C.; Smith, C.; et al. Brainstem blood brain barrier disruption using focused ultrasound: A demonstration of feasibility and enhanced doxorubicin delivery. *J. Control. Release* **2018**, *281*, 29–41. [CrossRef] [PubMed]
30. Focused Ultrasound Foundation. Focused Ultrasound for Diffuse Intrinsic Pontine Glioma (DIPG) [White Paper]. Available online: https://www.fusfoundation.org/content/images/pdf/Focused_Ultrasound_Foundation_DIPG_Workshop_White_Paper_2021.pdf (accessed on 13 May 2025).
31. Weisbrod, L.J.; Thiraviyam, A.; Vengoji, R.; Shonka, N.; Jain, M.; Ho, W.; Batra, S.K.; Salehi, A. Diffuse intrinsic pontine glioma (DIPG): A review of current and emerging treatment strategies. *Cancer Lett.* **2024**, *590*, 216876. [CrossRef]
32. Syed, H.R.; Kilburn, L.; Fonseca, A.; Nazarian, J.; Oluigbo, C.; Myseros, J.S.; Packer, R.J.; Keating, R.F. First-in-human sonodynamic therapy with ALA for pediatric diffuse intrinsic pontine glioma: A phase 1/2 study using low-intensity focused ultrasound: Technical communication. *J. Neuro-Oncol.* **2023**, *162*, 449–451. [CrossRef] [PubMed]
33. Dasgupta, P.; Balasubramanyian, V.; de Groot, J.F.; Majd, N.K. Preclinical Models of Low-Grade Gliomas. *Cancers* **2023**, *15*, 596. [CrossRef]
34. Mackay, A.; Burford, A.; Carvalho, D.; Izquierdo, E.; Fazal-Salom, J.; Taylor, K.R.; Bjerke, L.; Clarke, M.; Vinci, M.; Nandhabalan, M.; et al. Integrated Molecular Meta-Analysis of 1,000 Pediatric High-Grade and Diffuse Intrinsic Pontine Glioma. *Cancer Cell* **2017**, *32*, 520–537.e525. [CrossRef]
35. Roberts, J.W.; Powlovich, L.; Sheybani, N.; LeBlang, S. Focused ultrasound for the treatment of glioblastoma. *J. Neuro-Oncol.* **2022**, *157*, 237–247. [CrossRef]
36. Chen, K.T.; Chai, W.Y.; Lin, Y.J.; Lin, C.J.; Chen, P.Y.; Tsai, H.C.; Huang, C.Y.; Kuo, J.S.; Liu, H.L.; Wei, K.C. Neuronavigation-guided focused ultrasound for transcranial blood-brain barrier opening and immunostimulation in brain tumors. *Sci. Adv.* **2021**, *7*, eabd0772. [CrossRef]
37. Sarica, C.; Nankoo, J.F.; Fomenko, A.; Grippe, T.C.; Yamamoto, K.; Samuel, N.; Milano, V.; Vetkas, A.; Darmani, G.; Cizmeci, M.N.; et al. Human Studies of Transcranial Ultrasound neuromodulation: A systematic review of effectiveness and safety. *Brain Stimul.* **2022**, *15*, 737–746. [CrossRef]
38. Fry, W.J.; Barnard, J.W.; Fry, E.J.; Krumins, R.F.; Brennan, J.F. Ultrasonic lesions in the mammalian central nervous system. *Science* **1955**, *122*, 517–518. [CrossRef]
39. Krishna, V.; Sammartino, F.; Rezai, A. A Review of the Current Therapies, Challenges, and Future Directions of Transcranial Focused Ultrasound Technology: Advances in Diagnosis and Treatment. *JAMA Neurol.* **2018**, *75*, 246–254. [CrossRef] [PubMed]
40. Lu, V.M.; Niazi, T.N. Clinical Trials of Focused Ultrasound for Brain Tumors. *Cancers* **2025**, *17*, 513. [CrossRef] [PubMed]
41. Young, C.C.; Narsinh, K.H.; Chen, S.R.; Ansari, S.A.; Hetts, S.W.; Lang, F.F.; Wintermark, M.; Kan, P.T. State of Practice: A Report from the Inaugural SNIS Neurointerventional Oncology Summit. *Am. J. Neuroradiol.* **2025**, ajnr.A8902. [CrossRef] [PubMed]
42. McVeigh, L.; Patel, T.; Miclea, M.; Schwark, K.; Ajaero, D.; Momen, F.; Clausen, M.; Adam, T.; Aittaleb, R.; Wadden, J.; et al. Updates in Diagnostic Techniques and Experimental Therapies for Diffuse Intrinsic Pontine Glioma. *Cancers* **2025**, *17*, 931. [CrossRef]

MDPI AG
Grosspeteranlage 5
4052 Basel
Switzerland
Tel.: +41 61 683 77 34

Applied Sciences Editorial Office
E-mail: applsci@mdpi.com
www.mdpi.com/journal/applsci

MDPI

www.ingramcontent.com/pod product compliance
Lightning Source LLC
LaVergne TN
LVHW070448061126
851858LV00011B/190